THE NUNEX METHOD

A 7-Step Guide for Documenting Technical Projects Successfully

Written By: Richard Nuñez
Edited By: John Palmisano & Eric Shapiro

iUniverse, Inc.
New York   Lincoln   Shanghai

The NuneX Method
A 7-Step Guide for Documenting Technical Projects Successfully

iUniverse books may be ordered through booksellers or by contacting:

iUniverse
2021 Pine Lake Road, Suite 100
Lincoln, NE 68512
www.iuniverse.com
1-800-Authors (1-800-288-4677)

ISBN: 0-595-32585-8 (Pbk)
ISBN: 0-595-66649-3 (Cloth)

Printed in the United States of America

For my wife and daughter, Zulma and Clarissa
"My reason for reasons"

CONTENTS

TESTIMONIALS AND REVIEWS

The NuneX method could be called "The successful engineer's survival Guide." Rich Nunez provides the methodology, concepts and tools for engineers who wish to excel in their field. Let's face it, the demanding roles of IT professionals often lead to less than adequate documentation. Whether you are a seasoned engineer or just starting out, following the NuneX principles will significantly improve the quality of your work. I particularly enjoyed the real world scenarios encountered by Mr. Nunez. In addition to the structured documentation procedures, the book contains numerous examples of best practices that are normally learned the hard way. Reading the book, it becomes very apparent Rich has invested many years in the field and the compilation of this resource. This book is a must read for any IT professional.

Luckily, I've had the opportunity of working with Rich Nunez on many projects. I was able to see firsthand the results achieved by the implementation of these procedures. I consider Rich to be a friend, peer and mentor. I am confident you will enjoy this book and benefit greatly by following The NuneX Method.

David Ryan, CCIE # 6996

"Mr. Nunez has brought his first-hand experience as a technical IT professional to bear on the number one challenge IT personnel face today: managing an ever-increasing workload in a fast-paced, challenging environment. While covering real world topics like security concerns, expense reporting and communications in a highly readable manner, the author puts forward a framework for documenting IT work that will be of great help to those who strive to be successful in this arena."

David Kaiser, Project Management Professional

Fun to read—great real world examples!! You make all techs out there feel that they're not alone with the problems they have. You teach "tech-smarts," which all techs need.

You deliver on the expectations set forth at the beginning of the book.

It's an all-encompassing technician's "how to" masterpiece.

You cover the whole gamut-from fixing a printer to managing an enterprise-wide project. It's a must read for any IT Professional.

This book should be required reading by those enrolled in technical institutes/ colleges.

As I read, I saw how a software aid/tool to this book would make this method even better and easier to use/remember. I guess any tech could put it on their Palm Pilot…

You energized me to try the "NuneX Method"!

Steve Rundle, Systems Analyst

First and foremost please allow me to say thanks. It was a privilege to have read your book. I am not the best with words, so what I have done is to give you feed-back on the different chapters of you book. I will wrap it up with an overall overview of the entire book.

Introduction

When I was first asked to read the book, I honestly thought that it was going to be a book of procedures, with nothing to really grasp my attention. I found that the introduction really grasped my attention, and made me feel for you in that situation. I have myself been in similar scenarios, and as such this portion of the book made me feel very interested in reading the rest.

Chapter 1

Pre-Documentation has always been a part of my personal process, however I never realized it. This in fact may still be a weakness to my documentation proce-dures. One of the items that I found very useful was the contact sheet. I can count numerous occasions in which I needed to get a hold of one of the sponsors, and spent 20 valuable minutes searching for his contact info. This book could have saved valuable time and allowed me to concentrate and focus on the job at hand. Researching the material prior to going onsite for a job is very critical in my opinion, and I really like the detailed information that you provide.

Chapter 2

As a technician grows and matures into an engineer, he will find that having contingency plans becomes one of the most important aspects to the success of the engineer. A tech may damage a printer: "$300"; an engineer may damage a business: "$300,000." The stress levels change when you look at it in this manner.

I have found that planning my work upfront has greatly improved my performance and productivity. I started thinking that if I spent some time upfront, as you mentioned, I would have spent less time on actual hands-on work. Usually the hands-on stuff is work in the middle of the night or the weekend, so I really focused on improving these aspects of my game.

I have used the rules of planning to give me a checklist of things I may need to check or items that may have missed.

Chapter 3

I found that the security portion of the book was very interesting.

Chapter 4

Taking notes in my opinion is very important when dealing with certain types of technical operations. Installing a software application or operating system may require many steps. It would be very important to document this, if in case you need to rebuild it in the future.

Chapter 5

The worksheets that were available in this book allow for quick and easy templates to use with just about any project. I do believe that there may be times when these sheets will not be used due the nature of a project. Overall, my feel of this chapter is that it does indicate that documentation is a must. I personally document almost all my work; however, I do not have a systematic way of doing so. This may help with future documentation…only time will tell.

Chapter 6

The refining stage of the process is another important portion of the documentation process.

In my current professional career it an absolute must, and allows for faster and timelier responses to certain situations. It is also true that it leaves a footprint or a mark of a job well done. Customer Satisfaction is a very difficult thing to measure. However, I believe that good documentation will give very positive closure to any project.

Chapter 7

Quite honestly, this is where my documentation skills start to really get bad. I have found that I get closure on a project once everything is working and the customer signs off. If I need to go back for a certain upgrade or change in certain items, I don't document it as I would during the initial phases of a project.

Overall Rating of the Book "NuneX Method"

I may be a bit biased, having worked with you. Rich, I have had some experiences that really drove me to excel in documentation, which I would not have had unless I met you.

One experience was when we installed a web server for the Gadsden School District. I saw you take out your template and start making a plan of what we were going to do, and more importantly what was going to be the end result. I would have typically jumped right in and started working on the install, but I thought I would see your style of workmanship. All I can say is that from that day forward, I realized that having a game plan prior to engaging the situation at hand saves time.

We accomplished the implementation in less time than I alone would have done it in. I was impressed.

This book flows very well and gives you a process by which to start doing documentation. It does not however feel like a process.

Thanks again for the opportunity to read this book.

Sincerely,

Carlos Martinez, CCNA, CCNP, CCIE

Note: Carlos Martinez also utilized the NuneX and credits the application of the 80/20 rule as a key in his later passing the Cisco Certified Internetwork Expert (or CCIE) lab exam.

Rich, I must say you are brilliant! I think I might call your book the "IT Bible." This is great and I will always be thankful to you for providing us with this remarkable book of IT knowledge. In other words, "YOU ARE DA MAN!"

Marquita Calderazzo, Systems Analyst

Rich,

Congratulations! I am so excited about your book! I read the first two chapters, and it got me thinking about what an influence you had on my work style. About five minutes before I started reading, I finished a build on a dietary pathway that has taken me weeks to get working correctly, and was reviewing my documentation. Yes, I was reviewing my documentation, because I'd kept notes throughout the process. I'm always surprised that others are surprised that I keep documentation as I work and adjust it as I change things. I can't tell you how many times it's saved me time and rework to have detailed notes not only on what I did, but on what I planned to do. It's so easy to get pulled away from your task, only to return and not know where you were, or where you were going. It's just as easy to go down a wrong path and not be able to turn back because you're not sure how you got there. I attribute that part of my work process to your influence, and am thankful I learned the lesson early in my career.

I can also attest to the veracity of your statement that good documentation can get you farther in your career than you might expect. Customers, supervisors, and peers all value good documentation; though I don't always remember the details of a project after a few days, I can always go back to my notes or procedures for anything I might have forgotten. Not only that, someone else (ideally, with little effort) can read my documentation and get the same result—sure makes me look like a hero.

I often wish that other people did a better job documenting their work—what a boon it would be to myself and my customers not to have to redo work because others' efforts weren't well communicated. I'll highly recommend your book and techniques to anybody working in IT in any manner, from techs to consultants, and anyone expecting to work in IT upon graduation. It's great to see emphasis on documentation in IT; other industries like healthcare and real estate have been focused on it for years, and it's time we did the same.

Again, thanks for the impact you've had on my work style and success, and your willingness to share your ideas and offer help. You did a wonderful job communicating both the mechanics of your techniques, and relating them to everyday life. All of the tidbits that I picked up in a year of working for you are cohesively bound and easy to understand—a boon for those who have not had the opportunity to work with you. I look forward to reading the rest of your book (and any more to come) and refining my technique!

Sincerely,

Marisa Yater, IT Consultant

PREFACE

The purpose of this book is to instruct and encourage those in the Information Technology (IT) field who are seeking a better and proven method for documenting their technical work as it pertains to hardware, software, network troubleshooting, and project implementations. This guide book is primarily aimed at those who specialize in systems support, in the areas of personal computer (PC) and/or server systems as well as those who design, maintain, implement, and troubleshoot networked systems. This innovative methodology may also be utilized by the other specialty fields within Information Technology, such as system analysis, programming, and database administration, and by any IT specialist that may be called upon to technically work with systems at any degree or capacity. Some of the steps in the *NuneX Method* may not be applicable to non-technical IT specialists, but the majority of these steps will be useful for documenting your technical work even if you do not routinely perform technical tasks. It is also a beneficial methodology to serve as a foundation to create a documentation standard that may be required by your IT shop or organization. If your organization already has documentation policies in place, these will supersede the *NuneX Method's* seven steps of technical documentation as presented in this book and other related materials. Our goal is to minimize and eliminate your liability or culpability for violating company policies. In cases where your company's documentation policies conflict with the *NuneX Method*, your company's policies and procedures should always take precedence. Keep in mind that the *NuneX Method* can serve as a documentation standard where there is none available—either individually, on a company-specific basis, or as applicable to a large organization.

Who should read this book?

Essentially everyone who works in IT as a help desk specialist, technical resource, pc technician, network administrator, network engineer, project manager, network manager, department manager, director, or CIO/CTO should read this book. The book's technical level is designed to address new entrants to the IT career field as well as seasoned professionals. This book is unique in its approach to project management for the IT professional because it addresses the reality of commonly doing technical work in an unstructured and hurried manner. Many project management books focus solely on the concepts and approach to project management informing their readers of the importance of documenting, yet they don't clearly inform their audience on "how to" properly document their respective projects. This book shows you how to document in a manner that will allow anyone to follow your work and build upon it.

One of the many complaints heard about IT technical professionals is the lack of quality documentation performed on the job due to the fast-paced work environment and lack of enforcement of documentation policies or standards. Often, documentation is usually done in a hurried manner. The *NuneX Method* is designed to help those who don't have any documentation standards to follow in the workplace (or within their own organization). These people wish to achieve success in their careers by being proactive rather than reactive, and the *NuneX Method* will be instrumental to attaining this goal. There is a need for better documentation as the IT field grows, and the existing standards may be too cumbersome or may lack the flexibility technical professionals need in order to produce good documentation of their work, projects, implementations, service calls, and repairs. The documentation techniques presented here can be used as a guide for training and for coverage—to promote project continuity as personnel are absent from duty due to vacation or sick leave, for instance. When one is moving on to other opportunities and wants to leave one's successors and employers in good shape, these techniques have the potential to be invaluable.

The theme throughout this book is focused on becoming (or enhancing your current career as) a successful IT professional. This information may also be of benefit to instructors, professors, IT Consultants, systems analysts, programmers, and software engineers. Anyone who is looking for a better and proven way to document their technical work is a good candidate to read about and utilize this flexible methodology.

This book is unique, the first technical resource of its kind on the subject of technical documentation for IT professionals who would like to enhance their abilities and improve upon their successes. It's a reality check given the flood of project management "how to" and certification books out on the market, because it provides a practical hands-on guide instead of a theorized and best practice dictation of compiled methodologies for the no-nonsense kind of technical resource. It helps establish the logical progression and bridge between the techie who needs to complete his or her daily assignments in the best professional manner possible and the seasoned project manager who has further enhanced his or her understanding and experience by undergoing standardized and focused Project Management training.

How this book is organized

This book is divided into chapters that correlate to each step of the methodology, starting with Chapter 1, Step 1, "Pre-Documentation Procedures," and continuing through Chapter 7, Step 7, "Maintenance." Along with an introductory chapter, I've included actual real-world technical documentation worksheets (Appendix A) that the reader can use to reinforce the instruction and methods taught in the book. The last chapter in the book, "Putting IT All Together," demonstrates the combination of all the documentation steps to produce an effective and adaptable working documentation system for everyday use. There are various terms used throughout the book pertaining to information technology, project management, and technical disciplines—all accommodating a level of reader knowledge that has been assumed by the author. For an explanation or definition of these terms, please refer to the glossary at the end of the book.

I have designed this book to be a compilation of every best practice and real-world application I have researched, learned, and documented over the years I've been working in the IT field. I give you the foundation to start on and let you take it from there. Technical professionals are usually very busy individuals, with not much time for personal pursuits. I wanted this plan to be quickly understandable with regard to the vast knowledge and resources herein, while keeping in mind the busy lives most of us techies have.

ACKNOWLEDGEMENTS

I'd like to thank all of those who contributed to the ideas, development, and testing of the *NuneX Methodology*. Though many of those listed here may not realize it, their friendship, discussions, training, influence, and ideas have helped me on my continuing journey toward developing and refining this methodology. First of all I give to thanks to Jehovah God, the Grand Creator for allowing me the joy to learn, explore, and experience the greatness of His creation and learn of his wonderful purposes for humankind.

Dennis A. Kramer and Christopher Goulette, two valued friends who introduced me to the world of computers at age twelve. Jeff Nichols, who gave me my first job as a Computer Technician, even though at the time I lacked real-world experience. Steve Hughins, who showed me the best troubleshooting techniques at the start of my career that I still use today. Harry Rowlings, who showed me you don't have to know everything to be the best. Robert F. Naylor, who taught me discipline, control, and attention to detail in doing technical work. To the excellent teachers at Fabens Independent School District in Fabens, Texas, who had such a positive impact on my education and upbringing. All the technical instructors at Western Technical Institute in El Paso, Texas, who taught me how to keep and update a technical journal. University of Texas at El Paso, El Paso Community College, New Mexico State University, and Doña Ana Branch Community College, for their great technology programs. To the curriculum program at University of Phoenix, which prepared me amply for public speaking and presentations. To Kennedy Western University, for allowing me the opportunity to start my pursuit of a Masters Degree in Management Information Technology. To TechRepublic.Com and The Gartner Group for being a great source of informative articles on Information Technology. To the great cities of El Paso, Texas, and Las Cruces, New Mexico, for their culture, technology, and education focus, and for their wonderful people.

Regarding those whom I have worked for, those whom I have worked alongside, and those whom have worked for me, the following people have contributed to my knowledge and success as an IT professional: Gabriel Marquez, Jerry Widmer, Brian Foster, Scott Bollig, Larry Saville, Jennifer Raife, Renard Johnson, Edgar "The Terrible" Chacon, "CD" Ron Herring, Michelle Contreras, Carrie

Spratford, Paul Renner, Hermann Masser, David Colon, Robert Paquette, Marisa Telles, Eddie Esqueda, Reymond Ramirez, Marquita Calderazzo, Daryn Widmer, Alfred King, Richard Park, Patrica Tejada, Manuel Verduzco, John Schink, Carlos Martinez, Efrain Cuellar, William Harvey Lindsay, Eddie Quispe, Carlos Gutierrez, David A. Jones, Richard "Dick" Holcomb, "Mighty Mouse" Richard Mariscal, Norman Barnes, Manny Aguilera, Kristopher J. Tapia, Brent Lawson, Joshua Rushing, Phillip Rocha, Agustin Lozano, David B. Ryan, Arthur Wilson, Reggie Candelaria, Juan Mares, Jorge Gomez, Phillip Garcia, Luis Prieto, Ronald W. Womack, Frank H. Trifilio, Tina Marie Womack, Brenda Marti, Tim Thulin, Eli Nevarez, Jose Carrera, Paul Lopez, Ricardo Milian, Raul Gasca, Rene Roque, Brian Ainsworth, Roy Lambert, Rick Cue, Jim McMillen, and Spencer Visconti.

A special thanks goes to "Crazy" Larry Coffey and Steve Rundle, for encouraging my work and following up with my progress on this production. And also Robert Burns, for providing me with a wealth of real-world examples and applications of documentation best practices.

I'd like to thank my parents for their support of my education and career pursuits: My mom for exposing me to the joy and love of writing. My dad for encouraging the start of my career in the technology field when I wasn't sure what I wanted to do with my life. To my grandmother Antonia, who was such an important part of my young life. To my brothers Raymond, Joshua, and Isaac. To my Uncle "Cippy," who taught me the joy of being a techie. To my family, who has always believed in and supported my career.

I'd like to thank my wife Zulma for putting up with me during the long hours I spent working on this project. And my daughter Clarissa for giving me the unconditional love, joy, and inspiration a father needs every day.

Finally, I'd like to thank all those not mentioned here that have both contributed to and benefited from the concepts of the *NuneX Method*.

inTꞀODUCꞋiON

It was the summer of 1994. I was in my early twenties, working as a computer technician for a computer services company in El Paso, Texas. I was assigned on-call duty for the weekend and I was enjoying my Saturday morning breakfast with my family. Suddenly, I received a page on my beeper. I recognized the number to be a movie theater client of ours that was located approximately 75 miles from where I lived. I called the number on the pager, talked to the duty manager, and soon thereafter determined that I was needed on-site for ticket printer support. I quickly gathered my tools and headed out to the site. It took me approximately an hour and a half to get there. As I drove, I thought about all the possible things that may have caused the reported ticket printer issue. I was not proficient in servicing these types of devices so I felt an uneasy nervousness about the assignment. My expertise at that time was in personal computers, local area networks, and operating systems, not printers.

It was the weekend of a major summer blockbuster movie release and there was a very long line of people waiting to purchase their movie tickets. As I approached the ticket counter I could see that only one of the two ticket printers was operational and that everyone in the ticket booth seemed stressed. I verified the printer problem and gained access into the cramped booth. The booth already contained two people who were working furiously, selling movie tickets and answering a single-line phone that was located between them. As I got inside the small booth I realized it was going to be a tight squeeze, and that I would have to work in an uncomfortable space on my knees. My impression of the people waiting in line to purchase their tickets was that they blamed me and my company for the ticket printer's failure, and I felt that they were looking at me with an "it's your fault the ticket printer doesn't work" stare. I felt added stress because I also had the theater employees asking me: "Is it fixed?" "Is it ready?" "How long will it take?" "It was working fine yesterday." And so on and so forth. I did everything I knew to try to fix the ticket printer, but no matter what I did it just didn't function properly.

After about 45 minutes of initial troubleshooting, I pulled out my support call sheet and decided to call my boss to request his assistance. After several failed attempts to contact him, I decided to call the manufacturer support line for this ticket printer. Keep in mind there was only one phone in the booth, which was constantly being used to answer questions from people who called inquiring about the movie schedule. I didn't have a cell phone with me to use. The phone in the booth was the only telephone available. I was able to successfully reach the printer manufacturer's technical support line and talk to the support engineer on duty. I had to talk on the phone, then get down on my knees to work on the printer because the line cord wasn't long enough for me to talk at the same time that I worked on the printer. I wrote everything the support technician told me to do in sequential order, along with the time I talked to him, his name, phone number, extension—any information I was given. After about an hour of trying different solutions and working in an uncomfortably cramped space, the support engineer said he was going to have to ship a new printer to the site. He informed me there was nothing else I could do except wait for the new printer to arrive and replace it. Since it was Saturday, the printer wouldn't be in until Monday morning. I asked if there was anybody else in his support team that might be able to provide any other solutions he may not have thought of, since this call was very urgent and affected the movie theater's business operation. He said he was the only one on duty that weekend and that he was certain this was the best resolution in light of all possible options. I wrote everything down on my service work order sheet. I then hung up the phone and explained the situation to the duty manager, along with what needed to be done. The duty manager was not very pleased and asked to talk to my supervisor, and I gave him my card and wrote my supervisor's name and phone number on it. I requested the duty manager's signature on my work order to provide confirmation of the work I performed at their site. He read the report and then signed it, requesting a copy for his records.

I felt I'd done the best I could, but the situation seemed hopeless and disappointing. I went straight home from the customer's site, feeling disappointed with the lack of my success in resolving the issue. The experience left me with a lack of confidence in my ability to work on and troubleshoot printers.

The following Monday morning my boss received several calls from the movie theater's management complaining about our service work during the weekend. They threatened to terminate our support contract if we didn't get the issue resolved that day. My boss did a great job of dealing with the customer, and keeping them from terminating our much-needed support contract. We received the new printer that morning and I personally delivered it to the theater and replaced their broken-down unit with the new one. I then shipped the broken unit back to the manufacturer.

All seemed to go well after the printer was replaced until a week later, when my boss received a call from the manufacturer regarding the broken unit that was shipped to them. After about a five-minute discussion, he hung up the phone, called me into his office, and from the look on his face, I could tell he wasn't happy. He told me the manufacturer's repair technician said the problem was a simple fix and that I should have been able to resolve the issue on the field and that it had not been necessary to send in the printer. On top of that, our company was going to be hit with a repair fee for the returned unit—and we would be responsible for all shipping charges. My heart started pounding like crazy as he expressed his frustrations with *me*, personally, because of this issue. He scolded me for what seemed almost five minutes before he gave me the chance to respond. All I can remember were the cuss words—how I was responsible for almost losing an important contract, and that this would certainly be grounds for termination of my employment.

I didn't know what to say or how to respond; it had all come upon me so quickly. I had never experienced being fired before, and I guessed that's what it felt like. As I struggled to form a coherent response, I suddenly remembered the documentation contained within the work order I had made, and respectfully asked him to review it before he took any disciplinary action. My boss read it over several times, and then, with a serious look on his face, he apologized to me. I felt a huge sense of relief, and my boss then faxed my documentation to the manufacturer and followed up with a phone call to the technician he'd spoken to earlier. After the technician reviewed my documentation he also apologized to my boss for the unwarranted accusations of bad service work and admitted the problem was internal to their company. Later, after the dust settled, my boss assured me that if I hadn't made the extra effort to document the service call as I had, I would've certainly been fired. It would have only been my word against the customer's and the manufacturer's support technician. Fortunately for me, the written word was mightier than the verbal sword.

After that experience, I have made it a point to document my technical work for critical projects, and as a result have seen my career successfully advance from that of a computer technician all the way up to executive management. I have not always been the most knowledgeable or experienced technical resource, but the quality of my work along with the documentation I produced have helped me achieve a greater degree of success than I could have ever imagined otherwise.

This book is a compilation of various technical documentation techniques and best practices that I have developed over time, as learned from my personal experiences and from others in the IT field. I've conducted interviews and observations of successful technical professionals and have researched the topic of technical documentation and its best practices. In all my efforts to learn, understand, and apply the years of knowledge, insight, and wisdom gained from this endeavor, I've found that at the heart of every successful IT professional comes the ability to plan their work, work their plan, and document it in a way others can follow.

As IT becomes more of an essential business function and the demands for an experienced and capable IT workforce increase, the need to invest heavily in documentation techniques and knowledge management are increasingly important. The prevention of lost time and profits (typically a result of employee turnover) are becoming a paramount concern for companies that rely on technology to achieve their business objectives. Turnover in IT occurs as IT professionals seek better employment opportunities in other organizations, and as a result of cutbacks—usually related to economic slow-down—leading to layoffs. Companies that provide superior monetary compensation and an environment that promotes success will recruit and retain more experienced and well-seasoned IT professionals. When describing the ideal profile of a successful IT professional (not excluding the importance of personality and work ethic) a description would primarily encompass: diligent effort in documenting individual work and the work of a team; the ability to start and finish projects with impressive results; a systematic approach to achieving desired outcomes; and the ability to continually learn and improve from experiences in the field.

The methodological system known as the *NuneX Method*—named after its developer, Richard Nuñez—has been developed over a period of 14 years of experience, documentation, and experimentation geared toward developing and refining a workable system that can deal with and remain flexible to the influx of progress and change within the Information Technology industry. The main goals of this system are for any technical professional to properly document a project, implementation, request, or repair and maintain a project library of their own for reference and professional growth. It can even serve as a gauge to measure the success achieved by an IT professional as improvements are made through exposure to new ideas and techniques. Use of the *NuneX Method* can certainly contribute to an IT professional's own personal success story and be a tool to utilize anytime, anywhere. The icing on the cake is that an IT professional developed it for IT professionals, namely those who work in technical areas. The methodology wasn't developed by a bunch of theorists who thought this was the best way to go. Instead, it was developed by real world IT professionals seeking a better way to achieve successful projects using known best practices and procedures.

The main goal of this publication is to contribute to your professional success as an IT resource by utilizing the *NuneX Method's* documentation techniques. Your successful use of these methods will determine my success as an educator, writer, and consultant. In short, your success is my success. The pursuit of success is a key theme throughout this publication and it is the foundation on which the development of the *NuneX Method* was based. The *NuneX Method* is similar to Stephen Covey's book *The 7 Habits of Highly Effective People* in the sense that it focuses on the seven steps of highly successful IT technical professionals. The

NuneX Method also serves as a form of technical knowledge management which comprehensively gathers, organizes, shares, and analyzes its information and documentation to further its aims.

This book is intended to provide a win-win methodology that will benefit technology services companies, their employees, their customers, and essentially anyone affected by its use. It is designed to help the reader achieve success as an IT professional by minimizing the liabilities that come with a project, reducing work-related stress, allowing the difficult task of handling multiple projects and deadlines, and ultimately maximizing work efficiency to promote professional growth and personal achievement. It will allow you to be in control of your projects and not let your projects be in control of you—in essence, working smarter and not harder. It is also important to note that this text **is not** a project management guide or a "how to write documentation for a particular audience" type of book. There are various other books available that cover these topics in detail. The *NuneX Method*'s primary audience is those technical professionals who work with Project Managers and professional documenters by providing them the feedback and project status updates necessary for formal communication. A lot of the concepts and steps within the *NuneX Method* overlap with these areas, but are primarily intended to be utilized by the technical individual as assigned project tasks are performed.

This method will help reduce costs associated with employee turnover due to the time-consuming process of having to bring new people up to speed on a project. The technical documentation techniques contained in this book enable standardization of information flow and continuity—how to pick up from where someone else left off.

The *NuneX Method* also serves as an important form of technical knowledge management. If a key technical team member leaves an organization, valuable project-specific knowledge and insight will not be entirely lost; these will remain within a project's documentation, and will benefit those who will continue and complete the work effort.

The *NuneX Method* is designed for individual and team input; anyone from a single technical source (such as a technician) to a larger group, or team (such as an implementation team comprised of many technical professionals), can employ these standards to achieve consistency and superior quality in their documentation efforts.

This methodology can also be used to train new recruits in the IT technical profession to achieve the experience level of those more seasoned professionals.

The *NuneX Method* can serve as an initiative within a continuous quality improvement system that seeks to maintain quality and accuracy throughout its documentation steps. It contains a feedback mechanism in its Refinement and Maintenance steps that can be regularly updated. The key elements of the methodology are the use of the 80/20 rule for efficiency, quality, and knowledge management.

Finally, the *NuneX Method* is based on the concept that it's not necessary to be the most qualified and knowledgeable person working a project to be the most successful. Basically, you don't need to know everything to be the best. Its key focus is the attainment of success within the IT career field by utilizing documentation techniques that often go unrefined and underdeveloped—even by top performers.

The *NuneX Method* offers a way to achieve success as an IT professional by applying these documentation techniques. However, before we can start with the systematic approach to documentation, we need to define what documentation is—and how it can help us achieve success in the Information Technology field.

Documentation—the key to a successful career in Information Technology

Documentation. The word triggers many thoughts in different individuals. Almost everyone in any IT career field would acknowledge its importance. It's often mentioned as one of the most important daily tasks in an IT professional's job. Despite the wide acceptance of documentation's importance, do many IT professionals really practice what they preach when it actually comes down to documentation? The sad truth is that documentation is usually done inconsistently—and if completed, is usually done in a hurried manner. Even though most IT organizations may have a standard for requiring documentation during all phases of a project, most feel a great need to improve their documentation procedures. Their efforts to document their technical work may lack the required quality to allow others to follow their progress.

In a survey study conducted from 2001 to 2002 by N-Corp Services LLC on their web-site, the following statistics revealed:

Of those IT Professionals Surveyed:

Regarding the *prevalence* of a documentation standard:
35% said their organization has a documentation standard for documenting
    technical work
33% said their organization doesn't have a standard at all
18% said they didn't know if their organization had a standard to go by
8% said they use their own personal documentation standard
7% said their organization has a standard but it is not enforced

Regarding *when* the need for documentation arises:
43% said they document during all phases of a project (beginning, middle, and end)
20% said they document during the end phase of a project
16% said they don't document at all
14% said they document their technical work during the beginning phase of a project
6% said they document during the middle phase of a project

Regarding the *improvement* of documentation:
62% said they felt they needed to improve their documentation procedures
27% said they felt they needed to *somewhat* improve their documentation procedures
11% said they felt they didn't need to improve

Regarding the *value* placed on documentation:
38% said that documenting their work was very important to them
31% said that documenting their work was important to them
17% said that documenting their work was somewhat important to them
8% said that documenting their work was the most important process to them
6% said that documenting their work was not important to them

Finally, regarding how much we *like* writing documentation:
33% said that they sometimes personally enjoy documenting their technical work
28% said that they personally enjoy documenting their technical work
20% said that they don't personally enjoy documenting their technical work
19% said that they personally enjoy documenting their technical work depending
    upon the project assigned

In summary, for most of us we can say:
- We recognize the need to document during all phases of a project.
- We feel we'd like to improve our documentation skills.
- Documentation is very important to us.
- We can even enjoy our technical documentation duties under the right conditions.

This book doesn't reinvent the wheel on the subject of documentation, but is meant to demonstrate that documentation doesn't have to be a laborious and boring task that one must do merely to do what is right and, meanwhile, maintain compliance with company policies and procedures. It is designed to show you an organized and proven approach to consistency in documentation while providing a method for you to learn from your experiences. The methodologies you will learn in this guide will make documentation a part of everything that you do as a technical resource interesting, will show ways to make it fun, and will ultimately contribute to your success as an IT professional.

What is documentation? Documentation is defined as the act or an instance of the supplying of documents or supporting reference or records. It is also the collation, synopsizing, and coding of printed material for future reference. The main idea here is information for reference. Is documentation best left only for the end of a project, task, or endeavor? The answer is *absolutely not*. Documentation's effectiveness comes from answering the main questions—who, what, where, when, why, and how—and by noting events or tasks before, during, and after. Only then can documentation serve as a true reference that can benefit all interested parties.

You may comment that "Documenting before, during, and after will take too much time. I'm already having problems meeting my deadlines." You may even feel that you are an unorganized individual who would have trouble writing down everything. In the next section I will share a personal experience of mine that proved to me that efficiently documenting before, during, and after a project results in greater returns and actually reduces the time invested; whether that be in a project, problem, or request. The reduction in time will allow you to pursue other opportunities such as spending more time with your family, seeking out more time to achieve certifications, and completing the often-ignored administrative paperwork most techies have a tendency to relegate to a lower priority, consequently causing our vital paperwork to lag behind.

You may wish to reflect on your own personal documentation standards as well as those of the organization you work for before you continue to the next section. Here are a few questions to ask regarding your personal documentation standards:

—*Do I follow my own or my organization's documentation standards every day?*

—*Are these standards easy to follow or are they cumbersome and boring?*

—*Will I achieve success within my career or within this organization if I follow these standards? Where will it get me?*

—*Do I enjoy doing documentation?*

—*Is documentation something I always do at the end of a project, or before, during, and after?*

—*Do the documentation procedures I follow build on themselves, or do I have to create new documentation items every time I do something—even if it is repetitive work?*

—How do I feel about the process of documenting?

—Do I have an organized approach in how I go about putting my documentation together?

Answering these questions will help set your mind on the topic of documentation techniques for technical IT professionals. Before we explore the topic of documentation further, let's first learn about an important business principle that applies itself to Information Technology and documentation: the 80/20 Rule.

The Pareto Principle (or 80/20 Rule), as applied to Information Technology

The Pareto Principle states that in every field of human endeavor 80 percent of all efforts produce only 20 percent of all results. Conversely, 80 percent of all results are produced by just 20 percent of all efforts. This is referred to as the 80/20 Rule or Pareto Principle. Back in the late 1800s an Italian economist and sociologist named Vilfredo Pareto (1848-1923) discovered this principle and applied it to business processes. His studies on the 80/20 Rule paved the way for a wide application of this principle to every field of human endeavor, including Information Technology. See Figure 1.

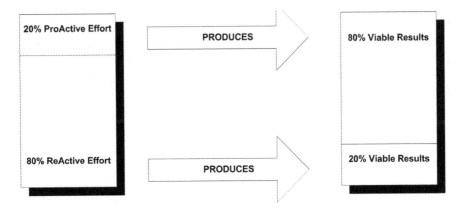

(FIGURE 1) 80/20 Rule or Pareto Principle

Another way to see the 80/20 Rule in a proactive or reactionary model is like this: 20 percent of your proactive inputs will produce 80 percent of the viable outputs. Conversely, 80 percent of reactive inputs will produce 20 percent of productive outputs.

The 80/20 Rule, if applied correctly in your work plan, can produce many opportunities for success and satisfying results. The goal here is to plan your work and *then* work your plan. In a technical capacity you should invest up to 20 percent of your time in planning and documenting your efforts so that 80 percent results can be realized. The goal here is to work smarter—not harder—by working efficiently and productively. The *NuneX Method* employs 20 percent of your efforts to produce 80 percent of the results by using the first four steps (Organizing, Planning, Security, and Notes) of the seven outlined in this book. The last three steps of the *NuneX Method* (Working Your Plan, Refining, and Maintenance Updates) produce the remaining 20 percent of results in a given project, resulting in 100 percent quality work!

You may tell yourself: "My work environment is too busy and fast-paced for me to allocate that much time on planning and documentation" or "I'm a hands-on type that needs to get busy and just jump into a problem and solve it." Some people think they do their best work when under pressure and tight schedules. I used to think this way until I analyzed my work efficiencies. If you really took the time to look at your work efficiency and measured it, you'd find that you are really not as effective under pressure as you could be. Working in a fast-paced manner often causes us to overlook quality and thoroughness, not to mention follow-through. The attributes of a job well done are typically sacrificed due to the fast-paced problem-solving you may be accustomed to doing.

Here's an example from one of my experiences dealing with Microsoft's popular Microsoft Office Suite's MS Word application. This example will show via my own personal experiences that fast-paced work is usually not the best means to go about resolving issues. This example is but one out of many I can cite. You too may have several examples that highlight the need to utilize the 80/20 Rule toward having a productive work experience.

Situation: I was sent to do some Microsoft NT server configurations at various remote sites, and I was scheduled to complete each site within one hour (the actual task took approximately 30 minutes to complete). I was doing some server configurations for the Marketing Department of a company I worked for, and while at the remote site one of the Marketing Department employees asked me to check out her MS Word application, which was doing very strange things. Wanting to exhibit the ideal customer service response, I told the customer, "This should be a quick and easy fix." Keep in mind that I had only one hour allocated to do my server configuration, and I needed to be at another jobsite to complete the next site's server configuration. I reasoned that I could take care of this request and then work on the server next.

Being familiar with Macro-viruses that often cause these types of errors, I decided to first run our McAffee anti-virus program to detect and remove the virus—which it did very quickly and successfully. Thinking I was the hero of the day, I then asked the customer to check the Word application again to see if all was well. As it turned out, her Word application didn't launch correctly anymore and a different system error message was being generated by the Word application. Thinking that the virus must have caused a corruption of some type, and not devoting more thought to the matter, I then automatically decided to re-install MS Word. "This shouldn't take that much time," I thought to myself, keeping track of the remaining 30 minutes of my pre-allocated hour. The re-install didn't work. Furious with my lack of results, I then decided to re-install the whole MS Office suite. Again, no success! I decided to un-install the complete MS Office suite and then reinstall the whole thing. No success. "What's going on here?" I asked myself, with only five minutes left in my allocated time. I felt really frustrated that a simple task had gotten so complicated, and now I was running out of time. As a result of my lack of success, the marketing person obviously doubted my ability to solve computer problems since I couldn't solve what seemed to be a simple MS Word problem. In essence, I wasn't able to meet my set objectives for that site.

I decided to tell the customer I had to be at another jobsite, but that I would return as soon as I could—or refer the call to another technician. Since she probably doubted my ability to solve the issue, she requested that it be referred to another technician. Feeling embarrassed about the whole situation, I called the MIS department and asked for one of my fellow techs take a look at the problem, and I went on to my next assignment.

As I went on to my next assignment, I was still feeling disappointed with myself for not being able to resolve what seemed to be a simple fix. This lack of concentration and harping on my emotional responses created a confidence problem that haunted me the rest of that day.

The next day, I returned to the client in the Marketing department to finish the task I'd gone there the day before to complete (the server upgrade). Much to my delight, I learned that the tech that had worked on the problem after me had the same trouble as I did. Furthermore, the other tech wasn't able to resolve the issue either. This made me feel better in a way, because now I had the chance to vindicate myself to the customer and to rebuild the confidence I had in myself as a service technician.

The server configuration task took me approximately 20 minutes to complete. I then decided to pull out my Microsoft TechNet CDs (an on-line knowledge-base of Microsoft Products issues, articles, problems, etc.) and did a query on the error message I was getting while launching the MS Word application. I received around 20 "hits" from the query, and as I surveyed the articles, I found one article that matched the exact scenario that I'd encountered on this particular system. I followed the resolution step-by-step (for all interested readers, the resolution was the need to rename the "normal.dot" file to "normal.old" and to re-launch MS Word, causing Word to automatically re-generate a new "normal.dot" file). The fix took only three minutes. I asked the customer to check for success, and was out of there within 30 minutes.

Lessons Learned:

Lesson 1: If I were to take care of priorities first (in this case the server configuration), I would probably not to have had to return to the jobsite the next day. I allowed myself to get distracted and lose focus of my initial assignment and objectives.

Lesson 2: As soon as I received the error message, I should have documented it in its exact phrasing. It would have taken some time to do this; however, the end would have justified the means. I could have faxed or e-mailed the error message to myself and to the tech who was going to work on the problem after me for research. I could have also called Microsoft Tech Support for assistance, since I didn't have the TechNet CDs with me during the initial call. I concluded I should always be prepared to handle the unexpected—and should always have my whole arsenal of tools and utilities with me.

Lesson 3: After noting the problem, I should have researched the error message on the Microsoft TechNet resource CD immediately, even though I felt sure some type of file corruption had occurred. Investing five to ten minutes in this would have saved the time that both the other tech and I ultimately spent on the issue. Sometimes we are too eager to leap before we look into a situation or problem. "Look before your leap" has always been good advice to follow.

Lesson 4: I should have documented my efforts on an on-line ticket system or even in a Help Desk log, so the next service tech wouldn't have to redo all the work that I had done if I wasn't able to resolve the issue.

Lesson 5: I should not have undertaken another assignment if I had a limited timeframe to work within at the site. In the spirit of customer service, I could have referred the initial call to our MIS Help Desk and provided them with as much information as possible. Taking on a quick assignment like this caused a level of stress and frustration that hindered my progress on the existing call *and* the next service call. I could have called the customer's site and asked if anyone needed any assistance prior to going there, and if all systems were operating without any customer concerns. This would have given me the opportunity to make adjustments to my allocated time as appropriate. I also should have been fully prepared for the assignment by having my copy of the MS TechNet CD with me.

Lesson 6: I should not have let the lack of resolution on the MS Word issue cause me to lose focus and concentration during my next assignment. This did nothing but make me feel inadequate, and to lose self-confidence for a brief period of time. A technician needs that sense of inner confidence to pursue his or her solutions—otherwise progress can be greatly hindered by mental obstacles. Feelings can affect one's ability to reason and use one's full mental faculties.

Overall lesson learned: If I were to have planned my work and worked my plan, I would have achieved 100 percent first call resolution, quality work, and excellent responsiveness to my customer's needs. If I were to invest 20 percent of proactive time to planning and documentation from the outset, I would have achieved the remaining 80 percent of viable results within an hour. Instead I spent 80 percent reactive inputs to only achieve 20 percent productive results. Both the other tech and I spent a total of 3.25 hours on the MS Word problem. I was not efficient, even though I worked at a fast pace with a hands-on troubleshooting approach. Also, quality work was initiated neither by me nor by the other tech by uninstalling and reinstalling the MS Office Suite repeatedly, which only resulted in losing the customer's preferred application settings (preferences). Even though this scenario was full of stress, pressure, and disappointing results, my reflection on the experience showcases a good example that "working on the fly" equates to *not* doing the best work possible. I know I am capable of a better work effort. This experience proved to me that working under pressure really didn't produce my best work effort; it produced a bad headache and an emotional roller coaster ride.

So what's your point?

My point is this: If you invest 20 percent of your allocated time in preparation (effective planning and documentation), you will achieve a high (80 percent) level of quality output. The *NuneX Method* is based on this same 80/20 Rule for efficiency, quality, and success.

In brief, here is the *NuneX 7-Step Method* for successful technical documentation techniques:

Step 1   **Pre-Documentation Procedures**
Step 2   **Planning Your Work**
Step 3   **Security**
Step 4   **Notes & Reference Information**
Step 5   **Documenting—Working Your Plan**
Step 6   **Refining Your Documentation**
Step 7   **Maintaining & Updating Your Documentation**

Each step will be covered in-depth in the following chapters. For now here's a quick synopsis of what each step concerns:

Step 1, "Pre-Documentation Procedures," involves setting up a Pre-Documentation checklist and organizing your approach, materials, and resources. It involves putting together a resource information database that contains names, addresses, phone lists, e-mail, web-site information, and so on. It also involves the psychological preparation that one needs to get started on a project. Technical people are usually brought in to resolve serious problems and are under an enormous amount of stress to find a resolution. Keeping focused and taking a balanced approach will help you achieve success at a faster rate. Giving documentation your first priority will help you achieve the success you desire in a faster, cleaner, and more productive way. One of the most important deliverables of documentation for any project is the record or accounting of your work efforts—and following the 80/20 Rule will help you keep things in perspective and achieve success.

Step 2 is "Planning Your Work"—in essence, conducting research, making preparations, and deciding which course of action to take in achieving project objectives. Effective contingency planning involves determining in advance what backup plans to use in case things don't go as planned, as well as leaving open the option of returning to where you originally started from in case things don't go as anticipated.

Step 3 is devoted to "Security" as it pertains to documenting your passwords and vital information. With the necessity of performing work on secured systems and networks, there is always a need for authorized access. Unless you are the one determining what the passwords are to be in advance (and this facet of the work should have a structured approach, too), you have to *uncompromisingly* document the system passwords you require to work on the equipment. The need for system access may arise as a result of providing support to a customer, an internal support call, or performing backups or system changes. Using a special encrypted coding method to securely document passwords for referencing system access is the major focus of this section. If your documentation falls into the wrong hands, you will want to be sure you've done all you can to protect your interests and deter hacks into the systems that you'd be held accountable for.

Step 4 describes the process of taking down "Notes and Vital Information." Notes are written information or downloaded documents, diagrams, excerpts from publications, etc.—any item that you use as a reference or for guidance in your efforts. This step also requires organizing these resources so that you don't have to look up the information as you work. Researching during a crisis is not a good task to be involved in, as oftentimes quality can be overlooked for the sake of timeliness.

Step 5 is "Working Your Plan" (the plan you made in **Step 2**). Follow your plan and implement your contingencies if necessary. This is where the actual documentation and step-by-step accounting of your work efforts take place. Utilization of documentation tools such as technical shorthand will ease the pain of laboriously writing everything down, especially when you are dealing with time constraints and pressure. You document everything as it happens—taking down all pertinent notes and observations. Technical notes and references used from **Step 4** are also noted here when applicable.

Step 6 involves "Refining Your Documentation," which was made during **Step 5**. After the dust settles and everything has been done, this is the opportunity to reflect back on your documentation (**Step 5**) and annotate the contributing factors to successes and failures that occurred en route to producing the desired objectives and deliverables. Basically, you'll go back and document what actually worked and what pitfalls are to be avoided the next time. This may be the final documentation product or deliverable you will provide to a customer or your employer, and it is *always* needed for your own records. You may also note the training and skill sets needed in order for another technician to assume the maintenance and support phase of the project. Knowledge management—encompassing the answers to the questions regarding who, what, when, where, why, and how—are crucial in this step.

Step 7 covers "Maintaining and Updating Your Documentation" after the project is completed or the resolution is reached. You will find it important to keep updating the results of your work, as sometimes problems and resolutions have a cause and effect relationship that may take time to reveal. Having a recorded accounting of events after you perform the project can help you refer back in case you need to perform clean-up work or analyze data for future reference.

This has been an overview of the seven steps to technical documentation techniques, or the *NuneX Method*. It may even bear some resemblance to the standard scientific method frequently used, which is basically:

1. Define the problem
2. Conduct research
3. State a hypothesis
4. Conduct an experiment
5. Collect Data
6. Produce a conclusion (whether or not the data supports the hypothesis)
7. Repeat the process (if the conclusion is in error or the data doesn't support your hypothesis)

In a way you can consider this a variation of the scientific method, fit for use by IT professionals. Following these steps will assist you in achieving 80 percent viable results. It will help you achieve the balance IT professionals often seek to help focus their time and energies on more important things, two examples being family and personal interests. Being able to achieve success on a daily basis and with every endeavor will add to your confidence and competence as an IT professional.

As you begin your journey in learning about the *NuneX Method*, I recommend that you obtain a standard three-ring binder, a hole-punch, and sectional page dividers. These are commonly available in most office environments and will serve you well in putting together a documentation binder. Please refer to Chapter 8 ("Putting IT All Together") for more specific information on putting together your system if you wish to start utilizing this system as you read, learn, and apply.

SUMMARY

Documenting effectively is best compared to a good physical exercise program: the more you do it, the better it gets. The resulting rewards will be that you will feel better about your work efforts because you will achieve some control over your work output. Following the *NuneX Method* of technical documentation will greatly contribute to your success as an IT professional, because you will have the added advantage of being proactive in an oftentimes reactionary industry.

CHAPTER 1

STEP 1—Pre-Documentation Procedures—Organizing Your Resources

Organizing your informational resources, performing required maintenance on resources, and updating project documentation after project completion are all vital phases of successfully documenting your endeavor. The first step is probably one of the most time-consuming of the seven steps of the *NuneX Method* because of the time it takes to acquire and catalog this necessary information. The payoff is the time it will save when you access all your available resources— quickly—to achieve success. Another advantage of Pre-Documentation procedures is that once you do the initial work, you are *very likely* to use the same information repeatedly. This is another demonstration of the 80/20 rule criteria, whereby 20 percent of your efforts in researching and organizing your resources produces 80 percent of the useable results. Further, you'll notice an increase of efficiency on your next project, as you are likely to use the same informational resources and people.

Here's the list of Pre-Documentation procedures to follow in order to achieve the first step of the *NuneX Method*:

A. Provide a reference name for the project you will be working on. Some naming conventions to follow are:

Code Abbreviation—Project (PRJ), implementation (IMP), service (SRV), support issue (SI), repair (REP), hardware (HW), software (SW), firmware (FW), maintenance (MAINT), installation (INST), upgrade (UPG), setup (SET), etc. These codes are useful for sorting electronic files quickly and easily identifying them for later retrieval. Dashes are used to separate codes, project names, and dates. A list of Code Abbreviations and their meanings is shown in Table 1. Please note that you may expand on this list or refer to N-Corp Services' web-site, found at http://www.n-corpservices.com, for updates.

Table 1: List of Project Code Abbreviations

Code	Description
INST	Installation
UPG	Upgrade
HW	Hardware
SW	Software
NET	Network
WAN	Wide Area Network
LAN	Local Area Network
DES	Design Project
SC	Support Call
PRB	Identify Problem Only
REP	Repair
SET	Setup
CON	Configuration
TRB	Troubleshoot
SRV	Service
SI	Support Issue
PRJ	Project (long term – typically over 1 month)
FW	Firmware
COM	Communications
FW	Follow-up
IMP	Implementation
MAINT	Maintenance

Project Name—This will be an abbreviated name for the project or task you are working on. You can use a period or dot to separate dissimilar criteria. Name the equipment, device, or software being worked on first and then separate the topic by using a period. You can also include a location or customer name next. The best approach in naming a project is to use the KISS principle (a.k.a. the "Keep It Simple" system).

Start Date—Date on which you began your project. One format to use is day/month/year. An example: June 3, 2004 would be 03June04. The most important facet of any date convention and usage is to *be consistent.*

Some examples using the **code abbreviation-name** and **project start date**:
   UPG-Mem.MainRtr-03Jun04 (a memory upgrade for the main router)
   INST-Win2003.Srvr.TimeClkApp-05Jun04 (a new Windows 2003 server installation for a time clock application)
   HWREP-PC.CIO-05Jun04 (a hardware repair of the CIO's personal computer)
   INST-MSExcel.10PCs@FIN-03Jun04 (installing Microsoft's Excel program on ten user PCs in finance & accounting—note the use of the @ symbol)
   UPG-CPU.Srvr@EPTX-08Jun04 (a CPU upgrade for your customer's server at the El Paso, TX site).
If a coded naming system is not your preference, name your project in a manner which will allow you to quickly and easily identify it long after you've completed it for easy recall.

B. Gather a list of all pertinent *internal* company resources: key personnel and their contact information. This will be a multi-purpose data sheet that you'll find yourself using often in your documentation. In this simple database, include phone numbers (cell, pager, residence, and office), e-mail addresses, chain of command (to whom an individual reports or their immediate supervisor—maybe include a copy of the organizational chart), duty hours, work and vacation schedules, on-call schedule, and home address (if allowed and available). Within your database be sure to note the date of the latest update, as personnel information can and often will change. A sample of this type of worksheet can be prepared using a spreadsheet or word processor. See the *Internal/External Reference Worksheet & Sample* (Appendix A) for a model of how you may put this information together.

C. Gather a list of the *external* resources that you usually work with (or will need to work with) in this particular project or assignment. As above, be sure to include phone numbers, e-mail addresses, chain of command information, and duty hours. Add other important data such as company name, website URL, account numbers, service codes, and which time zone they work in. See the *Internal/External Reference Worksheet & Sample* (Appendix A) for an example of how you may organize this information. You may prefer to combine both your internal and external contacts on one worksheet, as the sample demonstrates.

D. If applicable, contact your support vendors (in advance) to generate necessary service tickets or work orders, and advise them what project you will be working on. Be sure to reference this service ticket data on your external resource list for the project. Find out who will be "on-call," or try to get your hands on the on-call list in case you need to call after hours. Be sure to make arrangements for support in advance *if at all possible*. You may not need to call them in, but your support vendors can be prepared and willing to assist you if they know what issues to expect. This advanced notice may allow them to provide you any additional resources, ideas, or how-to's you may need to ensure a successful project.

E. Note any network addresses and subnets, acquire network diagrams (these may have been prepared by another resource), logical connectivity diagrams, software or database diagrams, interface connectivity guides, and standard (or default) settings for as many systems that you'll need or may come into contact with. If none of these diagrams are available, use the "divide and conquer" approach to create diagrams yourself, from your own technical perspective. A good method for network diagramming is to start from the system, go to the network connection, then the cabling to the network hub or switch where the connection may terminate, on to the network devices the system connects to, and finally the main hub or switch they connected to. Use any method or system that best works for you. You may elect to use the OSI (Open Systems Interconnect) reference model as your start and stop point—start at the physical layer and work your way up. If networking is not your forte, just diagram logically as much as you can for reference and understanding. For software and databases, ask a system's analyst, DBA, or vendor support for assistance in determining the logical layout of connectivity of the software or database structure. Find out if the application is multi-tiered, or functions in a client-server environment. Any information related to your project, no matter how insignificant, can later be considered the trump card toward success in your project or endeavor. Please refer to the following *Sample Network Diagram.*

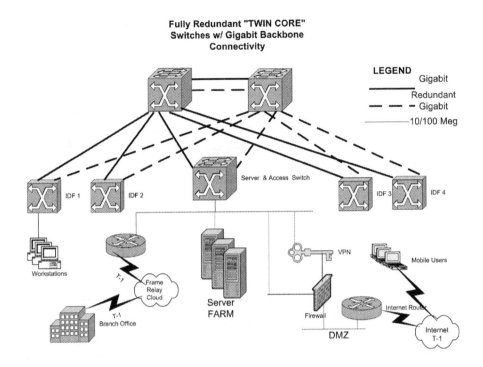

Fully Redundant "TWIN CORE"
Switches w/ Gigabit Backbone
Connectivity

F. Organize any tools, laptops, utilities, diskettes, CDs, installation CDs, programs, and license and activation codes which may possibly be needed so that progress is not hampered by lack of access to these resources. An example would be to take your licensed copy of Microsoft TechNet to work on a Microsoft product along with a separate laptop to run the application. Take a binder with CDs, diskettes, and your documentation worksheets organized for ready reference. Take a list of all license activation codes in case the need to reinstall an application arises.

G. Psychological self-preparation is the final checkpoint in the first step of the *NuneX Method.* Why is this important? Because being in the right frame of mind can enormously help your focus and awareness. Often it's the attention to this detail that makes the difference between a good job and an excellent one. A Star Wars Jedi creed states that "Your focus determines your reality." In essence, a clear focus on your work will determine your realized outcomes. As another popular saying goes, "Chance favors the prepared mind," meaning issues will be better handled if you're prepared to deal with them.

As IT professionals, we're often called upon to perform various tasks that take us on an emotional roller coaster ride—due largely to the stresses induced by deadlines and the complexity of unclear objectives. We need to prepare ourselves mentally to deal with these situations in order to achieve the desired outcomes.

Often what lingers in the back of our minds can create major roadblocks to successful effort. Examples of these could be personal matters, time constraints, unrelated concerns, and maybe even adversities. I often think about my family and the amount of time I wish to spend with them. At around 4:00 P.M. my wife will often call or page me to find out what time I'm coming home. Via her innocent inquiry, she inadvertently causes me stress by planting in my mind the desire to make it home as soon as I can. I may be in the middle of a repair or project and the loss of focus could throw me off track, or I may start making foolish mistakes by hurrying the project. In the back of my mind I am pressuring myself to get home as soon as possible, and as a result it *always* takes me longer to achieve the desired objectives. I have learned to force myself to re-focus and continue as if I was never interrupted. It takes time and practice to learn to deal with these types of mental obstacles positively.

I also try not to take on any non-priority project that may require two or more hours after 4:00 P.M., since I usually work from 8:00 A.M. to 5:00 P.M. I have to achieve this mental balance in order to perform at my best and achieve success. I try to save what I call the "no-brainer" or easy projects for the end of the day, so I can finish the workday on a positive note. Develop a system or routine that will enable you to take advantage of your peak hours, strengths, and energy, and that will allow you to have time to wind down the day in a positive manner.

Building your confidence

Numerous studies in modern medicine have theorized that there is a connection between mind and body, and that the state of either can and will affect the other. Motive and one's emotions can lead to either positive or negative reactions or consequences. There is no doubt that, as Information Technology professionals, we need to consider our emotional and physiological state of being—it can and will affect our work outputs. How can we achieve a good balance; how do we reach that place where our inputs produce productive outputs despite the external influences and strenuous situations that we may be exposed to on a daily basis? We need to build confidence in ourselves and in our abilities to handle various situations.

One way is through documenting and noting your reactions to various situations. You can theorize all you want about how you would react to a situation; however, until that particular situation arises—and it may differ from what you originally thought—you may react in a negative manner, and later regret your actions.

The IT world is full of projects, deadlines, problems, finger pointing, and personalities that may hinder your progress. These issues can create a stressful environment wherein one may not be adept at producing a quality output. How can we overcome these situations and lessen their impact on our work?

We can do this by applying a four-step process in our daily routines to help minimize the impact of unforeseen problems and any impulse for negative reactions to less-than-perfect situations. These four steps will help build your confidence to face the everyday non-technical problems and issues one may experience.

Four steps to building confidence
1. Preparation
2. Mentally rehearsing scenarios
3. Documenting your reactions, feelings, and observations
4. Reviewing and learning from your personal documentation

Preparation is the act or process of making oneself ready for a specific purpose, event, or occasion beforehand. Preparation can be seen as the 20 percent proactive input that produces 80 percent viable results. Taking the time to prepare yourself mentally, physically, and even emotionally will help you carry out your preconceived ideas and plans. Studying material or researching any related topics to help with the task at hand can be a good confidence-builder for doing your job effectively. If you know you are going to do a project in an area in which you lack experience, you can prepare yourself by investing time in the necessary training and research to be able to successfully carry it out.

Mentally rehearsing scenarios is the process of visualizing reactions—in a positive and productive manner—to various situations. A technique long used by track athletes in preparation for a race is to mentally rehearse the processes involved in running the race, long before the event takes place. This visual rehearsal is akin to a runner programming his internal system to react in the best way to the various stages of the race. Along with the physical training needed to carry out the plan (the race), the practice of mentally preparing to deal with the fears, anxieties, and stresses the runner may face has been observed to be a crucial part of training. With diligent preparation and practice, confidence to deal with these issues is increased. When the race actually takes place, a runner is like a batch program that is executed. Everything happens quickly, and the end result (if the program is executed correctly) is to be the first one at the finish line. IT professionals can train themselves to handle the everyday stresses and strains they usually face by mentally rehearsing scenarios. These stresses may include a difficult or irate customer, "problem" employees, short deadlines, reassignment, or the closure of a project that you may have invested a lot of time in. Regardless of the stresses induced (and there can be many variations), it's up to you and your knowledge of your own personal circumstances to mentally prepare for these situations. Like the runner that executes his program almost instinctively, you can quickly react to these problems in a confident manner and achieve personal success by the way you handle it. Practicing and rehearsing scenarios can lessen the impact of stress on your job and increase your confidence in dealing with these situations.

Documenting your reactions, feelings, and emotions is probably one of the hardest things for male IT professionals to do, since most men would describe themselves as logical thinkers and generally not as sensitive feelers. However, most professionals take pride in their work, and when work outputs are either positive or negative, emotions can instantly go on a roller-coaster ride of ups and downs, affecting work performance. Documenting your emotional state forces you to acknowledge any reactionary problems you may have, and can provide you with a good self-evaluation. You may be able to tell people in advance how you may react to certain issues, thus preparing them as well to help you handle it in the best way possible. By reviewing and learning from your documented reactions, feelings, and emotions, you can make adjustments to your methods or approaches to projects, problems, and issues. Reviewing and learning from your emotional reactions can later promote positive feedback. If you react by "stressing out," by having anxiety when a fast-track project is given to you (with not enough time for completion), and as a result you fail at the assigned task, you can later review your documentation and make the necessary adjustments. Maybe the next time (if the same scenario reoccurs) you can recall your response and choose to decline the assignment or lobby for more time and resources. If you don't reflect back to how you reacted and felt the last time, you may repeat the same mistake over and over again. Too many of these reactions may cause you to take extreme measures (such as looking for another job) because you feel you are being put under too many impossible deadlines—when it turns out you never really addressed and dealt with these issues appropriately to begin with. You may even repeat the same pattern at your next job if your emotional reactions aren't dealt with accordingly. You may never even know you have a problem dealing with various situations unless you take the time to document them. IT professionals work in a fast-paced environment where we may forget we experienced something until it is repeated.

These four steps can help you build confidence daily by making you think out your emotions before, during, and after each workday. The time spent on these issues can make you a resource everyone can count on and increase the value and commitment you bring to your IT shop. Confidence can help you tackle new responsibilities, tasks, and opportunities that you may not otherwise experience because of not being able to deal with the anxieties, fears, and similar responses that come with the job. Success is measured one project and endeavor at a time.

Summary

Step 1, "Pre-Documentation Procedures" involves setting up a Pre-Documentation checklist and organizing your approach, materials, and resources. It involves putting together a resource information database that contains names, addresses, phone lists, e-mail, web-site information, etc. It also involves the psychological preparation one needs to get started. Oftentimes technical people are brought in to resolve serious problems and are under an enormous amount of stress to find a resolution. Keeping focused and taking a balanced approach will help you achieve success at a faster rate. Building your confidence before, during, and after an assignment or task is key to achieving successful outcomes and improving yourself. Making documentation your first priority will help you achieve the success you desire in a faster, cleaner, and easier way. One of the most important deliverables of documentation is the record or accounting of your work efforts in a project, task, or repair. The mentality of the 80/20 rule will help you keep things in perspective and achieve success in all your endeavors.

Exercises (Answers are located on N-Corp Services' web-site http://www.n-corpservices.com under the downloads section)
In reference to Table 1, develop **code names** for the following projects (Hint: use the KISS principle):

1. Project—Personal computer memory hardware upgrade for 50 desktops.
Background: You work for a service company as a Computer Technician. You handle break/fix types of service calls, and you've been assigned a small project for an important account, the Smith Company. The project is to perform a personal computer memory hardware upgrade for 50 desktop systems, taking them from 128 Megabytes to 512 Megabytes of RAM. The manufacturer of the systems is Company X and all the desktops are the same model and have the same operating system.
Name of Project: _____

2. Project—Installation of a WIC module for a Cisco 3662 router.
Background: You work as a Network Administrator for a hospital that requires 24 X 7 up-time from its computer equipment. You've been assigned a project to add another clinic connection, Clinic A, to the existing main router. The T-1 circuits are not yet in place but you want to be prepared in advance—since you will only have a short window of opportunity when the circuits are ready. The main router is a Cisco 3662 that already has the necessary memory, firmware version, and

slots to accommodate the addition. The only thing you need to do is install a T-1 WIC (WAN interface card) module, and at a later time you will program the router and connect it to the previously-configured Clinic A router. You will need to power-down the router, install the WIC module, and turn it back on to perform the module insertion.

Name of Project: _____

3. Project—Network infrastructure design upgrade for Customer X.

Background: You are a Network Engineer working for a service company that specializes in network design, support, and implementations. You've been assigned to a long-term project for Customer X that encompasses the design and planning of a network infrastructure upgrade from an existing 100-Megabit backbone to Gigabit Ethernet. The design will involve the purchase of new equipment and re-termination of their existing fiber optic backbone to handle Gigabit specifications.

Name of Project: _____

4. Project—Troubleshoot the existing WAN network to determine the loss of connectivity.

Background: You are a Network Technician responsible for supporting remote sites within a 200-mile radius of the city in which your company is located. You are responsible for three sites: the Finance and Accounting site, the Education and Training site, and the Marketing Department's remote office. Each site has a LAN that connects to the corporate WAN with frame relay circuits ranging from 512K to a full T-1. You are only responsible for the LAN aspects of each site—the WAN support is done at your MIS Data Center by the Network Administrator. The most important site that you maintain is Finance and Accounting, since the CFO's office is located there and its business function is critical to the success of your company. Your work hours are 8:00 A.M. to 5:00 P.M. It's early on a Monday morning (around 7:15 A.M.), and you receive a page from the MIS Help Desk. It seems that the users in the finance site can't log into the network—or if they do, their connectivity goes up and down. You grab your documentation binder, head for the site, and start a new entry for this emergency project.

Name of Project: _____

5. Project—Create an internal/external reference list for your company and support vendors.

6. Project—Obtain (or create) a logical network diagram for your company (if you are authorized to do so). Be sure you understand all the connectivity points and what they represent. You may need to consult your company's network administrator for more information. After analyzing the diagram, answer the following questions:

    A. Is the diagram easy to follow and understand?

    B. How would you have documented the network differently?

    C. Does the diagram allow for at least 80% of your network to be rebuilt using the existing documentation?

7. Project—Obtain a written or on-line record of your IT department's documentation policies and procedures. Be certain to read it in its entirety. Look for answers to the following questions: Does the documentation policy address technical work or only system development or analysis? Is it enforced? If so, what are the penalties for noncompliance?

CHAPTER 2

STEP 2—Planning Your Work & Contingencies

Plan your work and work your plan—but recognize that plans may need to be revised when anticipated events don't occur. It's hard to do this when you are in the heat of battle, working toward solving an issue, or furiously trying to troubleshoot a critical problem. IT professionals battle with serious problems, projects, and issues every day. We also deal with limitations—in training, resources, time, and funding (to name a few). Regardless of the endeavor, one cannot ignore the value of planning, despite the urgency of most projects.

Planning is a major portion of the 20 percent proactive effort that will result in 80 percent viable outputs. Every plan should have at least two contingency plans in case the original plan doesn't (or can't) work. The *NuneX Method* employs various approaches and worksheets during the Planning phase. The overall project planning should be completed first, followed by contingency plans—assuming you already posses the basic skills and knowledge needed to carry out the work. If you lack the fundamental skills to carry out the project, you may opt to jump forward to the Notation and Research step of the *NuneX Method* first, obtain the necessary training for this project, and then return to the Planning phase.

In keeping with the planning sequence described above, the associated *Planning Worksheet* is to be completed first and the contingency worksheet next. If you fail to plan, your plans will most likely fail, and as a result you will not achieve the quality outcomes you desire.

Key questions to answer while planning are: *who, what, when, where, why,* and most importantly, *how.* I call these the "H5W Questions." To assist with these answers please refer to our *Planning Worksheet.* The worksheet has specific sections allocated to answer these questions in various sections. These sections specify the project name (*what*), the date (*when*), and the planners and project stakeholders (*who*). The Project Plan section answers the *how* question via the graphs on either the bottom or right-hand side. The graphs can visually serve as representations of *how* to accomplish the plan. The reason a graph is included is simple—a picture is worth a thousand words. If we can visually draw the plan, we can refer to it quickly and let it serve as a memory jogger of the plan we are going to execute. It can also help others follow our logical train of thought, especially if they are not adept with Technical Shorthand (later explained in chapter 5). The graph's grid lines can also be utilized for neatly drawing a flow chart or diagram for the plan. You can utilize a *Planning Worksheet* to prepare for a project and to get you thinking about the important aspects of a project, namely the scope, timelines, deliverables, sequence, results, etc.

PLANNING WORKSHEET

DATE: ___/___/200__ DAY _____ TIME: ___:___ AM PM PG ___ of ___ PGS											
Project Name:											
Planner(s):											
Projet Description:											
Projet Scope & Set Objectives											
Projet Deliverables											
Project Plan Sequence											

Plan Results	
Plan Completion Date: ___/___/200__ DAY _____ TIME ___:___ AM PM	

Do everyday technical tasks require planning? The answer is *yes*. Even though you may do the task every day, what if you're not there to perform the task? What if you're promoted—then who will perform these tasks for you? Do they know what is involved in getting the job done correctly? If you have a plan laid out and it works, anyone can follow your plan and achieve the same successful results. Your goal is to assure the same level of quality—or better—in your absence.

This chapter is the largest and most important section in this book, because it essentially covers the major portion of the 20 percent time allocation of the *NuneX Method*. Read it carefully to thoroughly familiarize yourself with the concepts.

Planning your success

What does it mean to plan your success? Basically every job, assignment, or project has a success measure by which it is judged. Is it simply good enough for us to complete the task and move on? Most of the time this is fine; however, to supersede your competition (both internally and externally) you need to let your success be known to all those whom you affect. A good success strategy measure to consider is this:

- **Completed work** requires notification that it has been done and sometimes gets acknowledged by the customer.
- **Good work** is seen only by those who do the same thing you do and often goes acknowledged by the customer.
- **Outstanding work** is seen, acknowledged, felt, and praised by everybody who is affected by your work (i.e., peers, management, customer, project stakeholders, etc.)

At which level do you wish to be recognized? I believe most of us want to provide outstanding service to our customers and organizations. In order to do this, you may want to set up success measures for projects as part of your planning. Ask your customers or stakeholders what they will clearly recognize as deliverables, and what meeting and exceeding their expectations call for. During the planning phase you'll consider the project scope, objectives, and deliverables, and how they will affect you. More importantly, you should also consider what will determine your success in your assignment and how to measure it. What accomplishments within the project must you achieve in order to perform outstanding work? One absolute to achieving outstanding work is by following the *NuneX Method* of technical documentation. The first four steps of the *NuneX 7-Step Method* alone will help you achieve 80 percent of the success measures you will be shooting for. The remaining

20 percent success measures are achieved by the last three steps. Achieving outstanding work may also involve other tangible things such as completing the project on time and below budget, and meeting the entire project's business needs and objectives as set by you and its stakeholders. Your project's successes are the tiny bricks that build the wall of accomplishment in your professional career, so don't ignore the value of planning your success measures.

One of the most convincing experiences I had regarding planning was when I attended a training session on a product I wasn't familiar with. I was sent to Colorado Springs, Colorado, for specialized training on a proprietary enterprise backup system. I somehow ended up in a group of people who were in the same situation as I. Most of the other technical people in the class were already familiar with and using the product in their companies. During the last day of training, we were split up into groups to work on a project exercise that would take several hours to complete. I was grouped with those that didn't know the product very well and would probably require assistance. I was developing key steps of the *NuneX Method* at that time and wondered if it would apply to training on a product I wasn't familiar with—and still achieve a successful implementation. This was my opportunity to test the Planning phase of the *NuneX Method* as it applied to first exposure to an enterprise level system without having the necessary background training and/or experience.

As we started discussing the project exercise, we noticed the other groups immediately started working on the system to implement the project requirements, since they were already experienced with the product. Most of the groups didn't organize their approach and just jumped in to the exercise. Some in my group felt uneasy about their rapid beginning, since we lacked the experience and confidence to use the product right away. Despite the uneasiness I was feeling, I somehow drifted into the role of group leader. I convinced the team that, due to our lack of experience with the product, we needed to thoroughly plan each step of the project requirements, research the needed information from our training manual to make sure our plan would work, lay out our plan in step-by-step detail, and finally execute it. We were given two hours to do this. We spent the first half-hour working on our plan, the second half-hour reviewing it and making adjustments, and the third half-hour working out the details of execution. By this time the instructor was getting nervous about what he perceived as a lack of progress on our part. Most of the other groups were almost done with the project assignment, but we hadn't even started typing a single command on the keyboard. Nonetheless, within the last half-hour we executed the project, in only ten minutes, and we followed our detailed plan to the letter. Everything went as planned and the project objectives were successfully met. To the instructor's surprise, we completed the project before most of the other groups. My assessment of the assignment was that we'd worked *smarter*, not harder.

Upon final review and critique of the exercise, the instructor shared with us the correct results for each project objective within the exercise. Our group had correctly answered 98 percent of the questions posed during the review while meeting all of the project's deliverables. The other groups weren't even close to the desired outcomes. Our thorough planning and execution (despite our lack of experience) resulted in better project outcomes than those reached by the other groups, which were far more experienced and familiar with the product than us. It was a great feeling of accomplishment for our whole team. Our levels of confidence and of product expertise were enhanced by taking this approach.

This experience helped to both motivate me and provide me with convincing proof that despite a perceived handicap of lack of product knowledge, one can still execute a successful project if thorough planning is conducted, checked, and then executed—in essence, planning your work and working your plan.

Successful leadership

Planning for success often involves taking a leadership role in order to get things done in a qualitative manner. As my experience demonstrated in the previous section by becoming a group leader in a training exercise, taking the lead role can make a significant difference regarding desired outcomes. In *You're the Greatest! How Validated Employees Can Impact Your Bottom Line*, the author, Francis MaGuire, mentions how successful leadership has three key ingredients: feelings, attitudes, and relationships. I like to refer to them as the "getting FAR (acronym) as a leader" absolutes. Successful leadership is not solely based on job knowledge, proficiency, earning potential, and the like, but rather on human qualities that can be developed, improved, and continuously worked on. Your passion and enthusiasm for your work and career (*feelings*), choice of outlook towards others and situations (*attitude*), and personal bonds, associations, and friendships (*relationships*) will help you become a successful leader in any environment if they exist in a positive manner. The validation of positive outcomes, good teamwork, and other noteworthy accomplishments is a must for taking a leadership role in the IT community.

IT planning rules

I'm sure most of you have an established methodology for how you go about doing system specifications and designs. What I'll present are concepts I have found to work for me (as well as people I've worked with closely) to achieve a successful project implementation. I've broken the list into a set of rules, meaning that they should be adhered to if possible. If I am able to do so, I choose not to participate in a major project that will affect my success measures unless the majority of these rules can be adhered to within reason. If you become too picky, you'll never be selected to participate in a project. Smaller projects are one thing, but for a major project assignment that affects my success measurement, I follow the rules of engagement for projects involving hardware, software, network products, and technologies. A prerequisite to using these planning rules is to thoroughly understand the business needs and objectives of a project (to the best of your abilities). *It is vital to note that without first addressing the business needs driving the project you are working on, the following planning rules will not be as beneficial and your successful completion will be without true purpose and meaning.*

Rule 1: Secure your area, systems, and networks.

Be sure to secure anything and everything related to your systems—especially the area in which your critical equipment is located. Make access to your equipment physically secure. Protection comes from establishing policies and procedures regarding use of the systems, setting password lengths, and using password generators, firewall, intrusion detection, packet filtering, and encryption technologies. Having a good security policy for administering your network is the first step toward securing your systems. Keep up-to-date on any new system hacks and patches for the Network Operating Systems that you're currently using so that security may not be easily compromised.

Rule 2: Protect your organization against harmful agents and viruses.

Invest available funding in anti-virus software that works with the system infrastructure you have and that is easily supported. Be sure to augment your investment with after-hour support, as many issues surface during non-business hours. Don't assign this task to someone who doesn't have a security background, lacks training in the product line you use, or may not possess a thorough understanding of the security requirements and responsibilities of your organization.

Rule 3: All systems (hardware, software, and network) must be fully redundant in every capacity.

If you can't afford redundancy for every possible system, then determine which systems would have the greatest impact on your business if you were to experience a catastrophic event—these systems should be the ones you make fully redundant. Hardware redundancy includes dual power supplies, dual controller cards, dual processors, dual NICs, and generally the use of mirroring for system partitions as well as appropriate RAID levels on data drives. Software must have a test and production system (if possible) on separate servers. All software configurations, customizations, and database work should be done on a test system first and then be thoroughly tested before the production system is changed. If your production system becomes inoperative you can then utilize the test system as a backup. Networks are recommended to follow the hierarchical model of network design, and allow for full redundancy by having (at a minimum) dual switched paths and at least two core switches. The key word is REDUNDANCY—in whatever way it applies to both you and to your vendor. You must strive to remove every single point of failure from your design and operation.

Rule 4: Plan for growth.

I've often seen a good design go bad by not planning for system growth. Since hardware prices have gone down over the years, growth can easily be factored into any project. A good rule of thumb is to plan for a minimum of three years and a maximum of five years of expansion space, and to assume from the outset that you may eventually replace or upgrade the system after three years of usage. Your cost savings today can be your space limitations tomorrow.

Rule 5: Make training an important part of your career and your team's career.

Training is one of the important things often overlooked because of the additional expense it adds to a project. It is often management's perception that a talented staff can quickly learn to do anything. However, the reality is that without the proper training and focus on a product or project you will not fully achieve success, quality, or the full potential of the system—despite the best efforts of most IT experienced professionals. I've often seen project personnel fail to reach their full potential due to the lack of training, even though the initial feeling was that success was possible without it. Good training accomplishes two important things: it places accountability on the trained staff and it increases everyone's confidence in their abilities to get the job done in a proficient and effective manner. A good standard is to always send two to three people to training if you can afford it. As a result of this buddy system you won't be at a total loss if one person leaves the team. You'll want to make sure you have training before, during, and after an implementation has taken place. As new people are introduced to the project it is wise to provide training on the product. One of the reasons management holds back from offering training to IT staff is the fear of trained staff leaving for better employment opportunities after having received the valuable training and experience with the given product or technology. Another reason management resists sending IT staff to training is the fear that the staff may not retain most of the information presented there. If expectations are established from the beginning (such as obtaining a certification for the particular product within two to three weeks after the conclusion of training), you can be sure that quality learning and application of the training will be achieved. Another notable practice is to require technical employees to sign a contractual agreement (if allowed in your state) stipulating that they will reimburse your company for any training provided to them if they leave. It is important to establish expectations and requirements upfront so that potential misunderstandings and legal issues are minimized.

Rule 6: Procure 24x7 support for hardware, software, and networks.

Even though your organization may only work from 8:00 A.M. to 5:00 P.M., it is a good idea to purchase round-the-clock support. Why? Because as your organization's IT technical resource, you're the one that is most likely going to have to work all night (or over the weekend) to get a system up and running by the next day. By maintaining 24-hour/7-days-a-week support, you have access to the additional resources you may need and won't have to worry about having to pay extra charges or outrageous rates for after-hours support. Even though you may have invested in training, you will not know it all and may lack expertise on a particular system. The best practice to follow is to try not to purchase anything critical to your business operation that doesn't have the option of an additional 24x7 support contract.

Rule 7: Have 100% capable backups, and test your disaster recovery plans.

Backups are often an afterthought to a project; however, when a system crashes, it becomes the main focus of the restoration. By the time a crash occurs it's usually too late for anything except to learn from mistakes already made. Invest in a good backup system with disaster recovery capabilities—and actually test it. Learn how to quickly restore critical systems and train your team members on how to be proficient in performing these vital tasks. If you have a fully redundant system, you may not need to do a full restore. However, not every organization can afford the luxury of a fully, 100 percent redundant system. A possible solution in this case would be to invest in a 100 percent reliable backup. Disaster recovery is a science in itself within the realms of IT. You may want to consider companies who specialize in disaster recovery services or have them provide you with consultation for updating your policies and procedures. A good disaster recovery plan would include the ability to restore the business functionality of an organization, along with the actual systems that are used to carry out the daily and critical business operations.

Rule 8: Document your operational processes.

It takes time and effort to learn how to perform a task, but once the task is learned proficiency is realized. What usually happens when someone new takes

over the task? The learning cycle is repeated and sacrifices in quality and production are usually the end result. If you have all your operational processes documented on a level where a literate child would be able to follow the instructions, then you can easily train someone else to run the same procedures. If twenty percent effort produces 80 percent results, then successful operations are usually guaranteed provided that effective documentation exists on these procedures. Document your processes and find ways to improve upon them. To paraphrase a popular saying: "If it ain't broken, then make it better!"

Rule 9: Have adequate power for future growth.
One of the most common mistakes in systems planning is failing to consider power requirements—especially for future growth and the addition of new systems. Backup power is usually an afterthought until a power outage occurs. Another issue related to this is system cooling—the environment of the equipment needs to be at a cool temperature to prevent overheating. Adequate power for immediate and future growth, as well as the ability to provide redundant power in case of an outage, are probably the best things to invest in to ensure that systems operate with minimal disruptions.

Rule 10: Monitor your systems—be proactive rather than reactive.
If your organization can afford it, it is generally a good idea to invest in additional personnel resources to monitor systems on a daily basis. Most IT shops can't afford this luxury, and must opt for the capability of system notifications to be sent to personnel via pager or some other means—after the fact. It is better to be proactive than reactive, since you will have time to do something about it with minimal impact on your business operations. Check often for security breaches and monitor for new viruses. Get updates from your HR office concerning personnel who leave the company—and don't forget to ask about their level of system access. Monitoring your system should also include keeping an **Issues Log** to track the history of the equipment, project, and implementation tasks that may cause problems later. Change control processes and notifications that are critical to maintaining a proactive response to issues that can surface later. There are many software packages available that specialize in this, such as Hewlett Packard's *HP OpenView* product suite, Computer Associates' *Unicenter*, etc.

Rule 11: Have a separate test and production system.
It is ideal and preferable to have a separate test and production system as a form of additional redundancy and protection from mistakes that usually are due to a lack of testing. Performing a controlled test in a test environment and appropriately documenting what occurs can produce better results when the fixes,

additions, and changes are later made to the production system. Organizations that can afford it normally have a build or staging system, a test system, a training system, and a production system. Since most organizations can't afford such a configuration, a test and production system should be the minimum criteria for a successful project rollout. Several virtual server technologies now exist that help offset the costs of running multiple test environments.

Rule 12: Standardize your tools and have the necessary tools & materials to complete the job.

The goal here is to standardize the use of the same troubleshooting tools, report writers, front-end development, and database engines (as well as back-end systems) as everyone else in your IT department or organization. Why? One reason for standardizing your tools is staff turnaround, and the other is to allow your staff to share ideas by providing an environment in which your teams can be synergistic. If only one person uses a tool, what happens when that person is not around or leaves, even though he has good documentation? Documentation is a reference resource, not a how-to-do-it-by-the-numbers instructional exercise. By providing the opportunity for everyone to know and use the same tools, documentation can be easily understood and followed.

It's advisable to have readily available tools and materials before undertaking the project or problem. Time is often wasted looking for simple things such as cables, connectors, torx screwdrivers, etc. Always have a small, commonly-used parts and tools package handy and ready to utilize. The five-minute fix should stay a five-minute fix because you're prepared to handle the small things—which can become "gotchas" in big projects.

Rule 13: Use well-supported technologies.

As new technologies are introduced to the IT industry (on an average of every three to six months) it is wise to avoid the "bleeding edge" that often causes grief. In many cases these technologies are so new that no one knows how to support them. It is therefore advisable to use technologies that have been standardized and supported by more than two manufacturers and/or vendors. I've often seen new products become reduced in price and in related support fees six months after being introduced to the market. The principle of supply and demand applies here: the less supply of a product, the more demand for product and support, resulting in price increases. This is due to not having many resources readily available to produce and support the product quickly. As a product becomes more available, the demand for support (along with product cost) decreases as more people become familiar with the product and as production is increased.

Rule 14: Manage licenses effectively, and realize the importance of auditing system hardware.

You may have followed all of the previous rules listed here, but if you ignore this rule you've set yourself up for failure—just ask anyone who has lost their job due to an unexpected software audit from those organizations that pursue software piracy, like the BSA (Business Software Alliance). It is imperative that you effectively manage any and all licensure related to hardware, software, and operating systems. A hardware audit or inventory can also be a great aid for predicting expansion and determining the amount of licenses to purchase at any given time.

Most Y2K projects produced good inventory counts. However, those organizations that maintain their inventory tracking are the ones that will achieve greater management and organization of their resources. If your organization ever gets audited and you were the one who ensured compliance, you will be the hero of the company because you'll have saved them thousands or even millions of dollars in fines. If your company failed to listen to your written and documented notifications of noncompliance, your liability is accordingly minimized and the blame can be placed elsewhere.

Rule 15: Hire competent staff.

Sometimes projects require the addition of human resources. When you have the task of hiring additional resources, it's a good opportunity to utilize the 80/20 Rule and select the best person for the job. If it takes you only one week to hire a person, and then two years to rid yourself of a bad hiring decision, then it would be wise to learn from your mistakes and make an earnest effort to find the best person for your team.

Take a lesson from the Disney Institute in hiring good support staff. I was fortunate to attend one of their many seminars in management training. What really got my attention (and is worth mentioning here) is that their hiring practices ensure excellent customer service. Their secret to hiring good "cast members" (as they call them) is by communicating their work culture to the candidate prior to hiring, stating their non-negotiable standards upfront, and creating a process of self-selection that is customer-friendly. Their self-selection process is very interesting in that it can weed out a lot of undesirable candidates from the outset.

If you're looking for a good, experienced network administrator you can phrase a statement within your advertisement for the position such as "We are looking for a well-experienced network administrator that can meet or exceed 99.99 percent system uptime requirements." Most unworthy network administrators would be discouraged by the 99.99 percent statement and would automatically *self-de-select* themselves. However, those who are seasoned, experienced professionals may feel they can achieve this, and will apply for the open position.

This small group of prospects who apply constitutes the people you want to consider. The next step would be to communicate your non-negotiable standards upfront. This could mean letting candidates know that the job requires 24 hour on-call time, weekend work, and coming in at odd hours on short notice—no exceptions. If you lay out the expectations and they still pursue the job, you can be more at ease with the quality of candidates.

When the time comes, have your best performers conduct the interviews. If you have a really good network administrator that consistently exceeds the job's expectations and responsibilities, you'll want this person to help select the new employee. Make sure you've offered him or her training in interviewing techniques, as inappropriate questioning could raise legal issues. A person already performing beyond the set expectations knows what it takes to deliver quality work and can recognize the same traits and work ethics in prospective candidates. On the other hand, they can also spot an impostor—someone who, in their professional opinion, may bring down the quality of your services and may pretend they possess skills and experiences they really can't prove.

Rule 16: Don't talk to salesmen unless you have money to spend. Don't waste time requesting quotes on products if you don't have the money to make a purchase, unless the purpose is budget preparation.

One of the most frustrating issues for IT managers is the continuous follow-ups of sales representatives wanting to close a deal. As an IT professional, you already have enough intrusions to deal with, and the last thing you want is an overzealous salesperson calling you during the middle of an outage. If you don't have the money, don't waste the time initiating contact or talking about your projects with salespeople. Look up pricing on the internet or other web portals that will provide you the information you need, and initiate discussions *only* when you have confirmation that you are ready to purchase (i.e., you have the budget, funding, or resources necessary to come through with a purchase order).

A key to sales is the ability to harvest and leverage relationships with customers. You need to be an unbiased customer, be able to put any friendships aside, and objectively consider all your options before you commit to accepting a solution from someone you like rather than someone who knows your business needs and is capable of meeting and exceeding your requirements.

Rule 17: Consider all possible options and be open to new ideas. Use your resource pool of staff, management, peers, vendors, research, etc. for help with planning decisions.

Nobody knows it all. However, a resourceful individual can find out what to do when the need arises. Don't get caught up in the arrogant reasoning that because you're the project lead and the most accountable person you therefore have the best and final answer to every question or issue. There is wisdom in having multiple counselors and viewpoints available. Consider all possible options and be ready for and open to new ideas.

Follow this brainstorming technique in planning sessions with your team of advisors:

1. Get your team together in a room where there are no distractions.
2. Set the rules for the meeting:
   —Any idea is welcome
   —No one person is the authority on the subject
   —There is no seniority to consider in submitting ideas
   —Write down all submitted ideas, even if they are repetitions
   —Don't permit criticism of any submitted ideas
   —Encourage participation by all attendees

The objective is to encourage idea flow and uninhibited creative thinking. You'll be surprised how good it feels to generate ideas as a group and not hold back due to fear of rejection or criticism. Finally, as a group, review the submissions and decide on the best ones to carry out.

Rule 18: Talk to the experts when it comes to product selections.
When you discuss product capabilities, request the advice of technical people who design, install, and support these products. It is preferable not to engage the salespersons who *promote*, but rather the personnel who *support* these products—since the salespersons' input may be biased in favor of closing a sale.

If you're a technician, it's best if you talk to those who speak your language. Unless the salesperson is technical, he or she may lead you to the wrong impression of the products or services he or she is trying to sell to you. The best arrangement is to talk to the technical resources that thoroughly know the product.

Rule 19: Keep the environment of the equipment room within ideal operating parameters.
Follow the manufacturer's specifications related to the proper cooling and humidity control of any area that may house vital equipment. Keep phone numbers and contact names for data center environmental control equipment representatives in case of an emergency (for example, keep contact information for a plumber in case you have a flooding emergency in the data center or equipment rooms).

Rule 20: Train the Help Desk to anticipate support calls regarding your project rollouts or go-live events.

One of the primary reasons implementations don't go smoothly is because the Help Desk staff typically is the last to know what is going on, and is usually involved only at the end of the project. Their involvement throughout each phase of the project can contribute greatly to a project's success. For instance, they may be the best source of answers to the *Frequently Asked Questions* pertaining to a project rollout, leaving you free to concentrate on more important issues. Work with your Help Desk crew to prepare them for upcoming rollouts, and they'll appreciate your efforts.

Rule 21: Be absolutely honest and don't hold back information from those who need to know, or who are accountable for the project's deliverables.

It's a human tendency to hide information that may be damaging or that has the potential for trouble. Honesty is always the best policy, and the more information you communicate (whether good news or bad) to those who truly need to know and are accountable for it, the better it will go for you in the long run. There is nothing more damaging to a project than for someone to know of an important issue, problem, or setback and for that information to not be communicated due to fear of reprisal or getting in trouble. It is better to get into trouble for doing something that is right than for something that is wrong. It may be difficult to do this at first because you may get your fellow team members in trouble. However, if you address their issues with them first and then escalate to their supervisors if not addressed, you will become known for providing fair warning to anyone that is holding something back.

Rule 22: Always take the time to personally thank all those who help you achieve success in your projects. Take care of those who take care of you.

People are usually remembered for the wrong things—when things don't go as planned or when blame needs to be placed. Remembering those who helped you achieve success is one of the best ways to thank them for a job well done. They will appreciate this acknowledgement, and will probably return the favor. Validation goes a long way toward building strong relationships within a team.

Rule 23: Plan your cabling needs.

My personal experience has shown that 80 percent of most network problems are attributed to cabling in some way. A bad connection, the wrong cable type, damaged cable, or a main distribution disconnect (and so on) are typical examples. If your cabling plan is well-conceived, well-designed, redundant, and documented, network communication issues should be minimized—if not eliminated. Cable problems are usually identified by a common network troubleshooting technique to start on the first layer of the OSI model (the physical layer) and to work up from there until the problem layer(s) is isolated.

Rule 24: Eliminate human dependencies.

If we as human beings are imperfect, the processes that depend on our abilities will also be imperfect. If possible, eliminate as much of the human dependencies as you can, striving for complete system automation and fail-safe operation. A system failure is typically due to poor programming or a flawed design. Remember an early computer programmer's saying: *"Garbage In = Garbage Out"*—the GIGO principle. Using a human-dependent system can create problems for you when you have staff turnover or a need for substitute coverage. The only way around this issue is if you have cross-trained all personnel and have all your processes well-documented—see Rule 8.

Rule 25: Don't waste your energy and time placing blame for failures; get the problem fixed and documented. Focus your energies on fixing and documenting and leave the finger-pointing to management.

Too much time and money is spent on placing blame for failed projects and seeking accountability. Even though it has its place, the act of finding fault is a waste of energy that can dramatically delay the completion of a project. It is best to leave that alone till the end of the project and focus your energies on getting the issues resolved. Leave the finger-pointing sessions and fault-finding to management. Unless you are asked to assist, it is best to complete the project and become its savior, rather than a judge.

Rule 26: Don't start on a project until you have verified that all equipment, materials, resources, shipping, peripherals, etc. have been checked in and are ready for implementation.

Most projects go uncompleted because a vital piece of equipment is missing or key resources are not properly scheduled. It is wise, for the sake of progress, to properly plan the deployment of equipment, tools, resources, software, etc. before undertaking the actual work. The minutes invested in making sure all things are accounted for can save days of frustration spent waiting for things to happen, or for parts to arrive. Check wall outlets, power distribution, cabling, infrastructure distribution, site planning, and space limitations of the project area. If you are traveling to a jobsite this rule is *very* important to follow. Be sure to verify that all necessary parts have arrived at the site or that they are packed with your gear. Take extra nuts and bolts, tie wraps, cable straps, testers, etc. as needed. The goal is to accomplish as much pre-work as possible so that you don't have any re-work.

Rule 27: Always be prepared to ship equipment back. Make note of how equipment was unpacked in the event you need to repack it and send it back.

If the hardware equipment for a project was specified and ordered by a non-technical person, you can be sure that something will be incorrect—either the wrong part or product was ordered. It is a smart practice to note how the equipment is packed so that you can be sure to repack it, if necessary. This will allow you to send back items on a money-back warranty while minimizing the restocking fees associated with ordering the wrong products. Be sure to quickly receive equipment and check it out. I've personally kept equipment in boxes for months and then unpacked them only to find out they are incomplete (or holding the wrong product), and have had a very difficult time getting them returned for credit. Most resellers will not accept product returns after 30 days or if they're not in the original packing materials.

Rule 28: Always factor additional costs prior to, during, and after a project.

The harsh reality is that most projects don't finish on time and within budget. Since the time factor is highly variable, the cost factor can come back to haunt you. I recommend that you factor additional or miscellaneous costs associated with a project—anywhere from 10 to 15 percent (or more). Even after a project is completed you'll find yourself still having to provide additional expenditures in labor and materials costs. Always be ready to plan for additional costs for at least three of the main phases of any project (such as planning, implementation, and maintenance). They usually come in the form of outside labor and support costs.

Rule 29: Estimate the time you think a project will take and multiply that by a justifiable multiplier that will work for you.

When we normally estimate the time we think it will take to complete a project, we often overlook an important fact—namely that our projects will rarely go as easily and smoothly as we think they will. Due to this fact, it is wise to multiply the time we think it will take by a factor of 2.5 (or another multiplier that works for you in your area of IT specialties). The 2.5 multiplier comes from my own personal experience in technical job estimations, where according to my actual documentation, I've found most projects have taken anywhere from 2.25X to 2.60X than I originally estimated. Upon interviewing and reviewing the documentation of other engineers and technicians who provided estimates, I found actual time factors to be about the same as mine (ranging from 2.4 to 2.8). If you can't rationally come up with a multiplier based on your company's actual data or your own documentation efforts, you may elect to use the 2.5 multiplier until you have enough data to apply a different one. If you think a project will take 20 hours, then by applying this formula you're really estimating it will take 50 hours (20 x 2.5=50). It is better to have too much than not enough time allocated for a project.

Rule 30: Plan your site surveys and avoid assumptions.

An effective approach to planning an implementation or product rollout is to perform a survey (preferably on-site) of all affected areas before the quoting process even starts—in spite of what information you may already have. Innumerable changes could have happened since your documentation regarding these sites was last updated. Be forewarned: Don't assume that your customer will be familiar with all the details of his site. I've been burned too many times by basing my plan of action on wrong or outdated information. My first rule as a technician was "Never Assume," because "Assumption is the mother of all screw-ups." Don't assume that what was done before you was done correctly—always retrace the steps beforehand. It is imperative that your walkthroughs be thought out in advance and not done in a hurried manner. If you have to do a last-minute walkthrough, then I suggest you take a video camera or even a digital camera for later review and detailed analysis. Prepare questions and checklists and research the products that may meet objectives beforehand, so that you can envision the new configurations and steps necessary to carry out the project as you survey each site.

Tools to take on surveys and project rollouts include power cord extensions, digital camera, video camera, notebook, pen, pencil, adapters, walkie-talkies, floor plans, maps, tape recorders, step ladders, etc.

One of the most common experiences one faces when surveying a job is the lack of detail in the notes of an engineer or project manager because he or she mentally recorded most of the experience. Usually another team member will be performing the implementation, and any visual information (in the form of pictures or video) can be instrumental in avoiding another site survey—and in preserving your customer's confidence in your company. Using a digital camera or a video camera, with the permission of the customer, is a good way to build an impression of attention to detail and quality of work in the mind of your customer. Presenting a slideshow to demonstrate "before and after" scenes of a jobsite can make the difference between a good job and an outstanding one.

Rule 31: Require a *Scope of Work* or *Statement of Work* proposal (along with a project plan) from any third party you engage for contract.

Any time you engage a third party for any type of services you should request from them a detailed *Scope of Work* document, with a project plan, to hold them accountable for services they will be providing. This keeps everyone honest and spells out deliverables in writing. Insist on weekly or bi-weekly status updates, or require a contractor to provide daily e-mail status updates, so you'll remain well-informed and able to update your project documentation.

Rule 32: Remember the relationship between quality, speed, and price.

When planning a project it is important to consider these three requirements: quality, speed, and price. In a perfect world, organizations would have all three. In the real world, it's best to follow the philosophy that two out of three is certainly achievable.

You can find quality products and services that have speed, but they will usually be pricey. You can find fast, inexpensive products and services, but you can't guarantee that they will be quality. You might find quality products and services that won't break your budget, but you're going to have to wait for it.

Here's a concise mathematical representation of the concept:

HIGH QUALITY + HIGH SPEED = HIGH PRICE
HIGH QUALITY + LOW PRICE = LOW SPEED
LOW PRICE + HIGH SPEED = LOW QUALITY

In a real-world situation you will very rarely be able to obtain all three, and will usually be able to only achieve two out of three. When you select two of the criteria, you will usually sacrifice the remaining one (i.e., if you want a system that has quality and speed, you will pay the price by spending more on this solution).

Most Valuable Rule: *Get it all in writing and keep it accessible to all interested parties.*

It is imperative (for legal reasons as well as in the spirit of obtaining good results) that any communication between you and vendors or any interested parties concerning any proposals, communications, issues, problems, etc. be communicated in writing as much as possible. If a conversation is performed in person or via conference call, take down notes and send back a summary or minutes of the conversation via fax or e-mail. If they agree with what you summarized, they will not question the information. However, if they disagree, request that they respond accordingly in writing. It's amazing what a piece of paper can do to reshape a bad situation and cause people to take issues more seriously. There is more work involved in doing this, but misunderstandings can be nullified by using documented text, letters, memos, etc.

Keep any proposals and quotations as part of your planning section. Any notes or summaries taken during the course of the project would most likely be reserved for the documentation step of the *NuneX Method*. Be sure to *always* make hard copies of any information that is submitted to management (or anyone else) for review, even if you have an electronic copy. Sometimes documents will get misfiled (or lost), and we sometimes accidentally delete critical files or forgetfully store them—and have difficulty retrieving them later.

These are only about thirty plus rules that I've discovered and learned over the course of my many years planning implementations and projects. I'm sure other experienced IT professionals would make some other suggestions. If you have any planning rules that I've missed and that you could add to these, I welcome the input. Be sure to send me your rules of engagement via an e-mail to info@n-corpservices.com or by using N-Corp Services' web-site feedback form at http://www.n-corpservices.com.

Planning Your Travel

Most field engineers are required to do some type of traveling, be it in the local area, to another city, or even to another country. Regardless of where you may be required to go, you should make an earnest effort to plan your travel arrangements in advance, especially if you're traveling internationally.

The first consideration is what means of transportation you are going to use, whether this will be your personal auto, rental car or van, airplane, boat, train, or a combination thereof. Be sure to know how to get to the travel stations from where you will be departing and what times to get there—leaving plenty of room to accommodate last-minute changes.

The second objective is to know the location of your destination site. Know how to get there (or have a detailed map made) and try to anticipate any confusing routes you may encounter. Track your mileage using popular travel sheets for odometer readings, or use Internet maps that calculate the mileage between addresses when submitting your mileage reimbursement forms. If your company doesn't provide you these reimbursements, you may be able to utilize them yourself for tax purposes. There is nothing worse than needlessly stressing over how to get to a jobsite and then back home. You'll want your travel to be as easy and worry-free as possible, so you can concentrate on the tasks you were sent to do.

Helpful Travel Tips for IT Professionals:

1. Acquire flight schedules and customer service/reservation phone numbers for the airline you will be flying, print them, and put them in the *Pre-Documentation* section of your project documentation binder. Be sure to print the departure-to-destination itinerary (and vice-versa). You'll have your flight information in advance and will be able to change flights if necessary. Print out your itinerary and have at least two copies—one to carry with you, and one for your documentation binder. Place the itinerary in the *Documentation* section of your binder. Program your cell phone with speed dial numbers of the contacts, site main, home office, your supervisors, team, airline, hotel, car rental, etc. for easy and quick access.

2. If scheduling your flight on-line or by phone (and assuming you have a flexible travel budget) be sure to obtain tickets that allow you to modify your flight plans without any additional cost. Do this especially if you anticipate running into problems at the site you are traveling to, or if you expect delays in traffic en route to the airport.

3. Take an extra set of clothes, shoes, and personal items in the event that you stay an extra day or spill something onto your clothes and need to change. Be sure to pack personal hygiene items. If you are attending a conference or training program, be ready to bring training materials back with you. These can be heavy and bulky, so be sure to take an extra tote-bag or leave space in your luggage for these materials.

4. Using popular map software that is available commercially or on-line, print out the driving directions from the airport to your hotel, from the hotel to your jobsite(s), from your jobsite(s) to your hotel, and finally from your hotel to the airport. Take note of the drive times involved and major routes. Assign a designated navigator if traveling as a group to help prevent getting lost. Place these maps in the *Planning* section of your documentation binder. An easier means of accomplishing this with modern technology is to rent a car with a Global Positioning System (GPS) guidance unit, or to purchase GPS mapping software to run on a laptop or PDA. Simply enter the destination address, city, and state and the electronic GPS system will guide you to your destination.

5. You may have a lot of idle time waiting at the airport, so take your documentation binder with you to review your project plan and any pertinent details. Review or study the travel routes you'll follow to your destination. Be sure to learn major highways and cross streets.

6. If you have the contact information for the sites you are traveling to, be sure to call them in advance to request the best way to get there—and take good notes. Despite a map's details, some areas are hidden and hard to find without a guide to help you.

7. Take extra money, or at least one back-up credit card. Be prepared to deal with the unexpected—and keep all receipts for reimbursements. Obtain or use a frequent flyer card to get bonus miles when you travel via an airline, rental car, hotel, or any other agency you may use.

8. When planning your site schedules, be sure to give yourself plenty of time to depart and arrive at each site. Consider the possibilities of traffic delays, joining flights, adverse weather, etc.

9. Be sure to get plenty of sleep while traveling. Lack of sleep can reduce your concentration and affect the quality of your work.

10. Keep *all* receipts in a separate receipts envelope, and discipline yourself to put receipts there as you pay for products and services. Reconcile these receipts and complete your expense report on the last day of your trip, or as soon as you get back.

If you are traveling internationally for IT-specific work, there are a lot of things you should consider. Some of the most common are:

A. Have a US-issued passport.

B. Have various forms of photo ID.

C. Travel with traveler's checks instead of cash.

D. Obtain the appropriate permits for IT-related work in the specific country you are visiting.

E. Have a list of contact names handy, in case you find yourself in trouble.

F. Know where the nearest Consulate or Embassy office is located, and the phone numbers thereof.

G. Make sure your communication system, e-mail, phone, pager, etc. will function in the country you are visiting.

H. If you will be renting or driving a vehicle, obtain the appropriate insurance coverage.

I. Be aware of any international terrorism that may be happening at that location, and be ready to respond accordingly if you find yourself in a dangerous environment.

The following are safe travel tips from the U.S. State Department (from www.salary.com):

Before you leave...

- Read the consular information sheet for your destination country, which describes local conditions travelers can expect. Find out about any travel warnings or pertinent public announcements in case there are any perceived threats targeting U.S. citizens.

- Learn as much as you can about the local laws and customs of your destination through your local library, travel agents, and the Internet. When traveling, you will be under foreign jurisdiction and not protected by the U.S. Constitution.

- Make two photocopies each of your passport identification page, airline tickets, driver's license or state ID, traveler's checks' serial numbers, and credit cards. Bring one copy of each with you, packed in an area separate from your originals, and leave the other copy of each at home with a friend, relative, or coworker. Carry an extra set of passport photos to expedite the replacement process if your passport is lost or stolen.

- Leave a copy of your itinerary with your family, friends, or coworkers in case they need to contact you in an emergency.

- Bring a list of prescriptions you are taking, along with the actual prescriptions and generic names of the drugs. If you're taking medication that could be considered a narcotic, check the legality of the drug with a consular official and/or get a letter from your doctor stating your medical necessity. In some Middle Eastern countries, certain tranquilizers and amphetamines are illegal and possession can result in arrest.

- Schedule direct flights to cut down on time spent taking off and landing (historically the most unsafe part of flying). If you must take a connecting flight, avoid stops in high-risk airports or areas. If you feel uncomfortable flying to a particular destination, consider an alternative form of travel, such as by train.

- Travel light. Follow airline restrictions on baggage and be sure you can easily carry your belongings without help. Try to keep one hand free in case of emergency.

- Minimize your time in the public areas of airports. Move as quickly as possible from the ticket counter to the secure waiting area near your departure gate. Leave the airport as soon as possible on arrival.

- If you plan to stay in a foreign country for more than two weeks, travel to a remote area, or visit a country or area the State Department has deemed "high-risk," register with the local U.S. Consulate or embassy upon arrival. This way it will be easier to replace your passport or evacuate you in case of emergency.

When you get there...
- You are subject to the laws of the countries you visit and are not protected by the U.S. Constitution outside the United States. If you get into legal trouble, contact a consular officer immediately.
- Dress conservatively to avoid being a target for pickpockets and other unsavory types. Over-the-top haute couture or super-casual jeans and a tee-shirt can mark you as a tourist.
- Minimize the valuables you carry. Instead of cash, use credit cards and traveler's checks. However, keep enough cash on you for a cab ride back to the hotel, embassy, or consulate in case you get lost or in trouble. Lock your passport in your hotel's safe (don't leave anything valuable in your room) and carry a copy of the passport information page with you. If you have to carry your valuables, the safest place is in a money belt or pouch worn under your clothing.
- Wear the shoulder strap of your bag across your chest, with the bag away from the curb to avoid drive-by purse snatching.
- Walk purposely, and act like you know where you are going—even when you are lost. When possible, only ask directions from individuals with authority (police and military officers, hotel concierges, etc.).
- Learn a few phrases of the local language, or carry a phrase book with you in case you need to signal for help. Make a note of emergency telephone numbers, including your hotel, the nearest embassy or consulate, and the local police department.
- Be conscious of what topics you discuss with strangers—you could expose yourself as a target for crime or terrorism.
- Take photographs—but be careful. In many countries you can be harassed or detained for taking pictures of government facilities or military installations. When in doubt, ask permission before snapping a picture.

When you leave...
- Have all receipts for exported goods ready and in order. Travelers have been arrested for purchasing souvenirs that could be antiques or national treasures. Document your purchases, and secure the necessary permits if you are removing an authentic item from the country.

- Change your money. Some countries don't allow travelers' national currency beyond local borders. Check with your hotel concierge or the airport's currency exchange clerk for more information.
- Be courteous. Answer customs and security officers' questions honestly and be patient with long waits. After all, those specialists have a job to do, and part of it is keeping you and other travelers safe.

Planning your expenses & reimbursements

Along with planning your travels you must also plan on how to get reimbursed by your company (or the company for whom you are doing work) for expenditures you may incur while providing your services. Check on company policies and procedures on reimbursements and know what you can and cannot do in order to get your expenses reimbursed, then plan accordingly.

Know the process to get your expenses reimbursed. Some companies can take up to a month or more to pay your reimbursements, so be aware of any interest charges you may incur with your credit card. Try to use a charge account that doesn't begin accruing interest for at least 30 days.

Keep copies of receipts—as well as cash expenses—so that you may accurately report them and not have to worry about losing money during an onsite service call or project assignment. You may be required to resubmit your expense forms and receipts. As you receive reimbursements, be sure to reference these payments and account for what is owed back to you, as payments are often broken up and sometimes not paid in full all at once.

Planning your implementations

Planning your implementations can be like a limited form of project management. You'll track your timelines and associated costs, and schedule your resources. You *must* know where you are at any point in the project and be ready to adapt where necessary. Remember, most projects rarely deliver on time and within budget. Work in conjunction with the designated project manager and be up-to-date on the project you're a part of. Be ready to use any special report forms or formats to keep everyone up-to-date on your progress regarding your part of the overall project. You don't want people to be waiting for you.

Scope of work

A scope of work document defines the work involved in the definition, design, and production of the components of a project's deliverables. A scope of work document should contain a brief background description of the project and the main objective. *Scope creep* is the term often used to describe the continual extension of the scope of some projects, which often leads to a runaway project. A statement of the work content of a project is required to clarify the project scope. The work content of a project is the sum of all necessary tasks and activities required to complete the project scope in its entirety (its complete work breakdown structure).

An IT department that acts and operates like a service entity toward other departments (like a service company looking for business opportunities within the IT market) will achieve higher success with its intra-departmental projects and requests. The efforts made in this regard will help increase customer service ratings within your organization and minimize misunderstandings when performing intra-departmental projects and tasks.

Having worked as a Director of IT Services for a large corporate reseller, I can say that one of the key factors to our successful execution of a job assignment or project initiated by a customer is a scope of work document (also known as a *Statement of Work*) that clearly defines the objectives, deliverables, timeframes, project overview, pricing, and responsibilities of both the customer and the service entities. Misunderstandings and billing issues are usually reduced to a minimum when the standard procedures for customer engagements are carried out via a scope of work document, with an accompanying project plan.

Prior to working for a commercial company, I was the Director of MIS Technical Services at a hospital where I managed the technical services, help desk, and operations teams. When reflecting on my tenure there, I wish I had utilized the scope of work approach for the many job assignments and projects that we completed. I know that if I'd worked this approach, the rewards of a job well done would have exceeded everyone's expectations. My team always viewed the departments we served as our internal customers. However, the frequent misunderstandings that occurred due to scope creep always made it seem that the excellent customer service they received was of no accord. In a fast-paced organization, leadership and decision-making responsibilities often quickly change, and opportunities to have the wrong expectations regarding the deliverables and execution of a project can occur. The resulting confusion and frustration can even lead to intra-departmental conflicts.

I recall a particular project during my experience as a MIS Director in which we were relocating the Patient Accounting (PA) department's personal computers, network equipment, and peripherals from the main hospital area to a remote site. Their equipment consisted of terminals, printers, some PCs, one local file server, and the associated terminal servers used to provide connectivity to all their DTE devices. During the initial planning stages we were coordinating the move with the PA's director, management, and supervisors. The plan was just to do a relocation project—nothing major, except to set up a new router and associated network configurations for their remote site, for about 25 users. We estimated the job would not take us more than one week to implement, barring any unforeseen issues.

As we were within two weeks of the start of this relocation, the hospital hired a new Patient Accounting Director who drastically changed our project scope and deliverables to suit his personal expectations regarding the move. The new director wanted to replace all the terminals with new personal computers, add new printers with network capabilities, have us provide training to the users (since their systems would be changing from terminals to PCs), and add e-mail and Internet access to all of their systems. Along with these new design changes, the Director also wanted to add another section of the PA department to the relocation project, increasing the user base to about 48. We had initially planned on assigning three of our technical staff to complete the project—now we would need to add more people to the project.

Obviously the changes in scope required the purchasing of new hardware, software, licensing, network design changes, and reallocation of labor hours and staff to complete the project within the week allocated for implementation. I wished we'd initially set up some type of written agreement with the previous PA Director that we could have used to keep the project from expanding its scope and requirements, since I had very little staff and more projects were slated immediately following this implementation. Unfortunately, I was not asked for input on this scope change but was told that it must be implemented without question. When I mentioned previous agreements were made and planning had already taken place, I was asked to provide solid documentation and was only able to provide meeting summary e-mails and high level overview documents regarding the move—unfortunately, not enough to keep the project closer to its original scope.

As it turned out, we were able to meet the new project scope—with much pain to my department. The systems purchases came out of my budget, and we had to work long hours to make this relocation happen. We also had a lot of post-implementation support issues that delayed us on our next projects. It took us a long while to recover from these problems from a scheduling standpoint. We also suffered a customer satisfaction decline (mostly due to resource availability) during this project and our intra-departmental relationship with PA went from bad to worse.

It's said that "hindsight is 20/20." I wish I'd utilized the approach our commercial entity now uses to engage a customer for a job or project. I now know that the standard approach for engaging customers for their business—and executing the objectives based on a scope of work document—would have minimized the headaches of dealing with scope creep within an intradepartmental assignment. An added benefit would have been ultimately improving the customer service ratings of our MIS Department, as expectations would have been clearly met and communicated. It would also have provided me the necessary documentation and backup for negotiating the project in the midst of the changes in management.

For those who may not be familiar with the standard methodology of engaging a customer from a commercial IT Services perspective, here are 10 Steps outlining the basic approach for generating a scope of work:

1. Meet with the customer to determine the best solutions to his or her business needs, especially where you can be the most effective in resolving key issues to problems or fulfilling specific service requests.

2. Put together a proposal of what you can do to meet the customer's objectives. Pricing, labor estimations, products, timelines, etc. should be included in your proposal. Consulting your team of advisors is important to provide realistic deliverables, expectations, and timelines as well as any insights they can offer.

3. Present your proposal to the key decision-makers.

4. Upon the customer's acceptance of the proposal, put together a scope of work document detailing what will be done. Within this document define the business requirements, objectives, deliverables, timeframes, project overview, pricing, and responsibilities of both the customer and the IT department (role definition). Include project assumptions that define what is included and excluded, along with the definition of successful completion of the project's deliverables. Be sure to include a "change order" process and form to be filled out and approved by authorized personnel when an alteration or modification to the project or request is required. Be sure to require that all parties need to reach an agreement on any proposed changes.

5. If the project is a lengthy engagement, provide the customer with a general project plan. When the project kick-off meeting date has been set, review the project plan with your customer in order to develop an updated project plan that works for all parties.

6. Have the customer or department head approve and sign your scope of work, and distribute copies to key people for reference.

7. Have an internal meeting with your project team to review any guidelines and the project objectives. Assign accountability and responsibilities. Provide status updates often, and communicate with your customer as important milestones are achieved.

8. Execute the project according to the scope of work document, and update the project plan as necessary.

9. Document the project according to your department's documentation standards—or use the *NuneX Method.*

10. Complete the project and turn over related deliverables to the customer. Secure the customer's acknowledgement of project completion in writing by way of a signed acceptance document. Request the customer's feedback on your team's performance and handling of the project, and address any unforeseen issues or problems that may remain.

These are the basic ten steps a commercialized IT services organization may follow while engaged in a project to prevent scope creep and misunderstandings, to ensure completion of the project or request, to maintain a favorable relationship with the customer and ensure quality work, and most importantly, to bill the project accurately and secure a presence with the account.

Even if the initial decision-makers change roles or leave, you can complete the project according to plan and ensure that any misunderstandings that occur due to scope creep, change in leadership, time of implementation, or other issues are minimized. Adequately fulfilling your assignments will allow you to add another notch to your team's list of accomplished projects and satisfied customers.

Treating other departments within your organization as internal customers and operating your IT department from the perspective of a commercialized service organization will make the difference when it comes to achieving successful project completions, and will improve intra-departmental relations. Accountability and ownership of a project will be defined with the scope of work document, and as a result there will be less room for scope creep and misunderstandings. You'll notice that your customers will be impressed by your service methodologies, and they'll be more appreciative of your team's efforts to handle their requests. More importantly, amid the popular trend of outsourcing IT departments, your IT department can operate confidently as a successful commercial services organization.

Project plan

The project definition phase is initiated by the usual signing of an order form, letter of intent, or contract. This is when the project is born and the appointed
Project Manager (PM) becomes the main driver of its activities. Based on the
scope of work documentation, the Project Manager can compose his or her planning schedule and budget—the so-called *Project Plan*.

"To Do" lists

"To Do" lists are the catch-all notes that will keep you out of trouble—especially
if you make commitments to complete certain tasks by a specified date or time.
They can also serve as reminders, as you may get distracted or sidetracked by a
newly surfaced problem, issue, project, or request. How do you prioritize a "To
Do" list? Here's an easy and effective sequence to follow:

1. Make a list of the things you need to do.

Example:

> Specify equipment for project
> Obtain quotes
> Purchase equipment
> Install equipment
> Register support/warranty info.

2. Compare the first two items and place a check mark or an X beside the most
important task:

> X   Specify equipment for project
> Obtain quotes
> Purchase equipment
> Install equipment
> Register support/warranty info.

3. Compare the first item with the third item, place an X beside the most important task, and continue this pattern all the way to the last item. In the following
example, "Specify equipment for project" has proven to be the most important
step of the group:

XXXXX	Specify equipment for project
	Obtain quotes
	Purchase equipment
	Install equipment
	Register support/warranty info.

4. Compare the second item with the third item, place an X beside the most
important task, and continue the pattern all the way to the last item. "Obtain
quotes" has more priority than the remaining three items:

XXXXX	Specify equipment for project
XXXX	Obtain quotes
	Purchase equipment
	Install equipment
	Register support/warranty info.

5. Follow the same pattern to complete the priority list:

XXXXX	Specify equipment for project
XXXX	Obtain quotes
XXX	Purchase equipment
XX	Install equipment
X	Register support/warranty info.

6. After using this method, your final priority list will be as follows:

First Priority:	Specify equipment for project
Second Priority:	Obtain quotes
Third Priority:	Purchase equipment
Fourth Priority:	Install equipment
Fifth Priority:	Register support/warranty info.

Planning meetings

Are job tasks, assignments, projects, repairs, and success measures the only things you need to plan and prepare for? You may also want to consider meetings. Why? Meetings are the primary business channel for planning, dispensing information, and communicating issues. Do you ever find yourself wishing you were more prepared for a meeting and trying very hard to not sound uninformed or even unprepared? In such cases it would be advisable to invest the 20 percent in preparation and gain the 80 percent productive results in meetings by having your documentation binder with you. Having the documentation binder can help you have your information readily available for project status meetings as well as customer calls, meetings, or any situation where you might be called on to present information. Be sure to set meeting days consistently. One example would be to only have meetings on Tuesdays and Thursdays while reserving Mondays, Wednesdays, and Fridays for other work. You can dress up for your Tuesday and Thursday meetings and dress down, or casually, for Monday, Wednesday, and Friday (assuming your company policy permits it).

A good format to follow for facilitating technical meetings (unless you already have one) is shown below:

Conducting an effective technical meeting

1. Make sure you have all the necessary resources to conduct the meeting without distractions. Pick a quiet and non-distracting location to meet. Make sure you have prepared the meeting location by providing writing materials, overhead projectors, extra paper, a conference-call telephone with speakers, a whiteboard with markers, and flipper paper. Remember: 20 percent preparation achieves 80 percent effectiveness. Arrive at your meeting at least ten to fifteen minutes early to set up and test all the equipment. Be sure to prepare (in advance) the material or information for which you are responsible. Complete any pending tasks assigned to you prior to this meeting, if possible.

2. Start the meeting on time, even though all group members may not be present. This will demonstrate your level of seriousness, and the attendees who arrive on time won't be frustrated.

3. Set rules for the meeting. Set the "Do's and Don'ts" for dispensing information. Each meeting has a unique purpose and different needs. Set the tone for the meeting as positive yet serious, and affirm that the meeting is no place for personal attacks or placing blame—you want factual information to be shared, and you want to keep the atmosphere positive and informative.

4. Dispense any high level information meeting attendees need to know, such as funding issues, policy changes, or revisions. Review any previous meeting minutes that may need to be addressed. Pass out project information and assign responsibilities to those who need to be informed. Be sure to present *any* priority information first.

5. Give each person a time limit for presenting. If yours is a one-hour meeting and you have five attendees who will present, give each person (including yourself) the floor for up to ten minutes. Emphasize the need to pay attention, and also let attendees know meeting minutes will be taken. This will allow the group to focus their listening skills and not worry about taking down notes. You may also choose to record the meeting for later review. As a rule of thumb, attendees are to provide information on the following:

    A. Tasks previously worked on

    B. Tasks currently working on

    C. Next project or assignment

    D. Issues that are creating problems

    E. What the team can do to provide assistance

    F. Timelines

Keep track of the time. If it appears that someone may be dominating the meeting, be sure to kindly interrupt him or her and let others have their turns, then return to any person who may have more to say. There is nothing more frustrating than someone who dominates the meeting and doesn't let others express their concerns or impart their information.

6. At the end of the meeting, briefly ask what can be done next time to make the meeting more productive and beneficial for all.

7. End the meeting on time, and inform the group when minutes of the meeting will be circulated—and make it a priority to deliver on this promise. Share meeting minutes via interoffice mail, e-mail, or whatever medium your company uses. It can be frustrating to have made a lot of progress with a project, only to have someone new join the group and ask questions that were resolved weeks or even months prior. An alternative to explaining these resolved issues at new meetings is to provide previous meetings' minutes so a new person can get up to speed. Have new people address the team leader with questions prior to meetings so that other team members don't get discouraged.

These seven steps are very effective, and you'll see that they can be a great contribution to your documentation processes when you reach the notation phase of the *NuneX Method*. Please note the different worksheets available for planning your technical meetings in Appendix A.

Worksheets should be added to your documentation binder quickly. *Meeting Planning* worksheets go in the Planning Section. *Minutes* & *Attendees'* worksheets go in the Documentation Section. The meeting planning worksheet's form identification name relies on the same procedures as the project naming convention—except that it always begins with *MTG*, and ends with the month and year. Since your worksheet already includes the day, your meeting name does not need to include it.

Example: Project Status meeting for a network infrastructure roll-out can be named:
MTG-NetworkInfrastructure-MonthYear

MEETING PLANNING WORKSHEET

MTG Date: _____/_____/_____ Day: _____

Meeting Location: _____

MTG Facilitator(s): _____

_____

Will Attendees need directions to meeting location? _ Yes _ No

Needed Resources:

- _ Overhead Projector
- _ Overhead Computer Display
- _ Speakerphone Capability
- _ Agenda Sheet # _____
- _ Project Status Sheet # _____
- _ Network Access—Security _Y _ N
- _ Tape Recorder w/tapes

- _ Whiteboard w/markers
- _ Flipper Board w/markers
- _ Laptop Computer for Presenting
- _ Writing Materials (Pen & Paper)
- _ Attendance Sheet # _____
- _ Internet Access (Browser_____)
- _ Other: _____

Setup Time: _____ _ AM _ PM   Responsible_____
Start Time: _____ _ AM _ PM   End Time: _____ _ AM _ PM

Reference to Previous MTG Minutes Sheet #: _____

Change to MTG Format as designated by last MTG's Minutes

Meeting's Purpose

_____

MTG scheduled and all attendees notified? _ Yes _ No
Responsible _____

TECHNICAL PROJECT STATUS MEETING AGENDA WORKSHEET

MTG Date:_____/_____/_____ Day:_____

Meeting Location: _____

MTG Facilitator(s):_____

MTG Recorder/Secretary: _____
Start Time: _____ _ AM _ PM   End Time: _____ _ AM _ PM
Reference to MTG Planning Worksheet #:_____
MTG Rules _ New Set _ Addendum _ Revision

AGENDA

I Administrative
   _ A. Review of Previous MTG Minutes
   _ B. Policy Changes
   _ C. Funding Issues
   _ D. Project Status/Updates
   _ E. New Assignments
   _ Other _____

II   Team Member Status Updates
   _ A. Tasks previously worked on
   _ B. Tasks currently working on
   _ C. Next project or assignment
   _ D. Issues that are creating problems
   _ E. What the team can do to provide assistance
   _ F. Timelines
   _ Other _____

III   Improvements/Suggestions to MTG formats

IV   Meeting Minutes Updates Info.
   Dispensed via _ e-mail _ interoffice mail _ postal mail _ fax _ network
   _ Other _____ _ all of the above
   _ Date: _____
   _ Responsible: _____

V   Conclusions/Review/Adjourn Meeting

Contingency planning

Having a contingency (or backup) plan in place is like having insurance. You may not need to use it. However, if you have a situation that requires it, your losses are nullified or minimized. If you plan your work and work your plan and it doesn't seem to go the way you anticipated, you may feel pressured to come up with a new resolution that may sacrifice quality output or not meet the exact needs of the project. This is where contingency planning comes in handy. You should have at least two backup plans in addition to your main plan—a "Plan B" and "Plan C" of sorts. You may argue that contingency planning for everything takes too much time. However, by formulating backup plans, you improve on your preparation efforts to handle the unforeseen and you can reduce the time required if a backup plan didn't exist. Sometimes a contingency plan's implementation will result in a better solution than the original plan could have. At other times, contingency plans can be reused—allowing for the 20 percent time spent initially on putting them together to provide 80 percent quick results—if you have a similar project requiring the same contingencies.

The best way to find a problem area early on is to make an earnest effort to look for it, and the first place to look is where problems most frequently happen. Another thing to consider is that contingency planning is not just a one-time, "etched-in-stone" occurrence; it may change as new issues develop or new information becomes available. It's a good idea to review your contingency plan when there is any change in circumstances, or when a major setback affects your work.

The contingency worksheet can help organize the backup plan in case things don't go as smoothly as expected. The *Contingency Planning Worksheet* has self-explanatory fields for entering vital information such as external and internal resources for the project. (You may refer to your *Pre-Documentation Worksheet* for this data.) The "other" field can be used to reference a ticket, case, or work order number. The worksheet has contingency "Plan A" and "Plan B" areas to meet the requirement for two (at a minimum) backup plans. Use the sheet to reference the resources back in your pre-documentation phase.

An important point to make here is that you may not even use the backup plans, or they may not even work for you if you start using them. Their purpose is to make you ***think*** about what-ifs and to help you organize your time, efforts, and resources. Backup plans can assist in avoiding the stress of having to contact people at the last moment or trying to locate information when you are already having problems. They also help you by forcing you to research solutions to potential problems that may occur. Lack of organization will only cause you undue stress—and could even stress the people you'll be contacting at the last minute. People dislike another's lack of preparation and tend to have a low opinion of those who don't prepare for an important task. You will gain valuable respect from your associates if you prepare yourself (and them) for what is about to take place. It almost becomes an automatic pilot program of good support as you enlist the help of your team for your projects. This type of preparation will help ensure the project's successful completion.

Your "Contingency Plan A" should be to initiate a call to technical support for assistance. Your "Contingency Plan B" should be to restore the system to its original state—prior to your changes. Your "Contingency Plan B" can also be a plan to take the system or project back to its most stable operating environment, if possible. Your "Plan C" (and so forth) can be any other contingency plan(s) you may have thought of that pertain(s) to your basic plan.

CONTINGENCY PLANNING WORKSHEET

DATE: _____/_____/200___   DAY _____   TIME: ___:___ AM PM   PG _____ of _____ PGS
Project Name:

EXTERNAL SUPPORT RESOURCES

Resource Name:
Title:
Company Name:
Phone:
Cell:
FAX:
Pager
e-mail
Customer #
Other:

Resource Name:
Title:
Company Name:
Phone:
Cell:
FAX:
Pager
e-mail
Customer #
Other:

INTERNAL SUPPORT RESOURCES

Resource Name:
Title:
Department Name:
Phone:
Cell:
FAX:
Pager
e-mail
Customer #
Other:

Resource Name:
Title:
Department Name:
Phone:
Cell:
FAX:
Pager
e-mail
Customer #
Other:

Backup Day	Sun	Mon	Tue	Wed	Thu	Fri	Sat	Other
Backup SET								
Backup Date								

CONTINGENCY PLAN A

CONTINGENCY PLAN B

Summary

Step 2 of the *NuneX Method* is planning your work—in essence, conducting research, making preparations, and deciding which course of action to take to achieve your objectives. Planning your success measures is the core building block for performing outstanding work and achieving quality results. A list of rules of engagement to use when planning hardware, software, and network implementations can help you design and implement the best systems possible. Planning is also a part of facilitating meetings, as well as gathering and dispensing project status information. Planning is of high importance in the *NuneX Method* for technical documentation. With a good plan in place, 80 percent productive results will be realized within your project implementations. Contingency planning involves determining in advance a backup plan (or plans) to use in case things don't go as intended. They are designed to keep you thinking about what-ifs and to be prepared to handle the many uncertainties that come with project rollouts.

Table 2: Rules of Engagement Worksheet
Best Practices to Consider for Planning Your Projects

Rule #	Check	Rule Summary
1		Secure your area, systems, & networks
2		Protect your organization against harmful agents & viruses
3		All systems should be fully redundant
4		Plan for growth
5		Plan for training
6		24 x 7 Support for HW, SW, & Networks
7		100% capable backups—test disaster recovery plans
8		Document your operational processes
9		Have adequate power for current & future growth
10		Monitor all of your systems—be proactive rather than reactive
11		Have a separate test & production system
12		Standardize your tools & materials needed to complete the job
13		Use well supported technologies
14		Effectively manage licenses & audit hardware
15		Hire competent staff—set high standards
16		Don't talk to salesmen unless you have money to spend or are preparing a budget
17		Consider all possible options & be open to new ideas
18		Talk to the experts regarding product selections
19		Keep the environment of the equipment within ideal operating parameters
20		Train the Help Desk for support calls regarding your projects
21		Don't hold back information from those who need to know
22		Always thank those who help you
23		Plan for cabling
24		Eliminate human dependencies
25		Don't place blame, get problems fixed
26		Don't start until everything is ready
27		Be prepared to ship equipment back
28		Factor in additional costs for a project
29		Multiply your estimated time for a project by a factor
30		Plan your site surveys & avoid assumptions
31		Require a Scope or Statement of Work proposal along with a project plan from any 3rd party you engage for contract
32		Quality, Speed, & Price
*****		Get everything in writing—all communications, correspondence, etc

Exercises (Answers and examples are located on N-Corp Services' web-site, http://www.n-corpservices.com, under the downloads section):

Please refer to the *Project Planning* and *Contingency Planning* worksheets for completing the exercises below. Please note that some information has been added or changed for each exercise in order to better cover the concepts presented in this chapter.

1. Project—Personal Computer memory hardware upgrade for 50 desktops. Develop a project and contingency plan using the provided *NuneX Method* worksheets.

Background: You work for a service company as a Computer Technician. You handle break/fix types of service calls, and you've been assigned a small project for an important account, the Smith Company. The project is to perform a personal computer memory hardware upgrade for 50 desktop systems, taking them from 128 Megabytes to 512 Megabytes of RAM. The manufacturer of the systems is Company X and all the desktops are the same model and have the same operating system. You must complete this within three days and your window of opportunity to work on these systems includes early morning, lunchtime, and after hours slots. The Smith Company's business hours are 8:00 A.M. to 6:00 P.M. The customer requires that the project is completed before their end-of-month processing begins.

2. Project—Installation of a WIC module for a Cisco 3662 router. Develop a project and contingency plan using the provided *NuneX Method* worksheets.

Background: You work as a Network Administrator for a hospital that requires 24/7 up-time from its computer equipment. You've been assigned a project to add another clinic connection, Clinic A, to the existing main router. The T-1 circuits are not yet in place but you want to be prepared in advance—since you'll only have a small window of opportunity when the circuits are ready. The main router is a Cisco 3662 that already has the necessary memory, firmware version, and slots to accommodate the addition. The only thing you need to do is install a T-1 WIC (WAN interface card) module, and at a later time you'll program the router and connect it to the previously-configured Clinic A router. You will need to power-down the router, install the WIC module, and turn it back on to perform the job. Clinic A is due to be connected and ready to operate 20 days from the day you've been given this assignment.

3. Project—Network infrastructure upgrade for Customer X. Worksheets include *Rules of Engagement, Project Design & Planning, Scope of Work,* **and** *Project Plan.*

Background: You are a Network Engineer working for a services company that specializes in network design, support, and implementations. You've been assigned to a long-term project for Customer X that encompasses the design and planning of a network infrastructure upgrade from their existing 100-Megabit backbone to 10 Gigabit Ethernet. The design will involve the purchase of new equipment and re-termination of their existing fiber optic backbone to handle Gigabit specifications. Use the *Rules of Engagement* worksheet to draw out information from the client on how to provide them the best solution for their investment. Develop an initial *Scope of Work* document and a *Project Plan* for your design.

4. Project—Troubleshoot the existing WAN network to determine the loss of connectivity. Develop a *Project Plan* **and** *Contingency Plan* **using the** *NuneX Method* **worksheets. Be sure to plan your travel and expenses for this project.**

Background: You are a Network Technician responsible for supporting remote sites within a 200-mile radius of the city in which your company is located. You have responsibility for three sites: the Finance and Accounting site, the Education and Training site, and the Marketing Department's remote office. Site 1 (Finance and Accounting) is located in City X, 45 miles from your corporate office. Site 2 (Education and Training) is located within your corporate office's main city—except it's in the far west side of town, about 25 miles from your corporate office. Site 3 (Marketing) is located near the corporate office, across the street in a business plaza. Each site has a LAN that connects to the corporate WAN with frame relay circuits ranging from 512K to a full T-1. You are only responsible for the LAN aspects of each site—the WAN support is done at your MIS Data Center by the Network Administrator. The most important site that you maintain is Finance and Accounting since the CFO's office is located there and its business functionality is vital to the success of your company. Your work hours are 8:00 A.M. to 5:00 P.M. It is an early Monday morning (around 7:15 AM) and you receive a page from the MIS Help Desk. This message explains that the users at Site 1 can't log into the network—or if they do, their connectivity goes up and down. You grab your documentation binder, head for the site, and start a new entry for this emergency project.

5. Project—Develop a planning and contingency planning worksheet for a new project that you are currently working on or that you've just started. Be sure to use your Internal/External Reference list from Chapter 1 as a reference. If you conduct meetings for this project, use the *Meeting Planning* worksheet to plan a project status meeting.

6. Project—Use the *Meeting Planning* worksheet to develop a meeting plan for the project in Exercise #3 (the network infrastructure upgrade project).

7. Project—Use the task prioritization technique covered in this chapter to develop a logical, sequential list of tasks needed to execute Exercise #1 (personal computer memory upgrade project for 50 desktops).

CHAPTER 3

STEP 3—Security

When I was in high school I was a very shy and private person. I respected the privacy of others, and I wanted them to respect my privacy. Unfortunately, adolescent teens are quick to reveal secrets and get involved in the personal affairs of other teens. Like most young males in their junior and senior years at my school, I had a girlfriend to whom I wrote private notes, and I'd deliver them to her between classes. Sometimes I used friends as carriers to deliver my secret messages. I knew that teachers discouraged this type of behavior among their students and often confiscated the notes and read them out loud in class to purposely mock and discourage this type of activity. Rightly so, they wanted their students to focus on the things being taught in school and not on their own personal adventures. Knowing that the information I wrote to her in these notes may eventually fall into the wrong hands (teachers, nosey students, etc.), I devised a coded system for my girlfriend and I to use. It took us several days to learn it and use it effectively, and before we knew it, we were writing encrypted coded messages to each other every other class. Sometimes teachers would catch us reading our secret notes and confiscate them, but since they didn't know our coded system, they would simply either hand the note back to us or tell us not to do that anymore and discard it in the trash. It was a great means of securing our privacy and preventing others from reading our daily communication, especially if teachers wanted to read our notes aloud in class. It was a great feeling, knowing we'd found a way to beat the system of prying eyes.

As I've grown up and progressed in my education and career, I still use a coded system. However, this time I use it for business and securing passwords and information. This coded system is the basis of Step 3 of the *NuneX Method*—Security. As the Internet, political, and corporate atmospheres of today's information age depend on Information Technology to achieve their goals and objectives, security has become a prime factor for consideration. It is a common saying among IT professionals when describing Security that "if someone wants to access your systems bad enough, it is inevitable that they will eventually find a way to break in." It's also an everyday reality that as a technical IT professional you will need to gain access to users' PCs, account names, passwords, and administrator accounts. You will also need access to network devices such as routers, switches, hubs, terminal servers, etc. Sometimes it takes a while to accomplish your work because you lack the correct access information or someone feels uneasy about giving it to you. Even when you obtain the access information, it's only normal to quickly dispose of the piece of paper or mental note containing vital access information because you don't want to be held accountable for any future security problems.

The first rule of network security: Create a policy

The key to strong network security is implementing a security policy that sets standards, assigns responsibilities, and puts mechanisms in place for staying up-to-date.

Password handling best practices

The following are considered by many IT professionals to be the best practices for handling passwords within an organization. This information was obtained from TechRepubic.com, from the article, *Make a Password Policy Part of Your Security Plan,* by Rick Vanover:

- No passwords are to be spoken, written, e-mailed, hinted at, shared, or in any way known to anyone other than the user involved. This includes supervisors and personal assistants.
- No passwords are to be shared in order to cover for someone out of the office. Contact IT, and it will gladly create a temporary account if there are resources you need to access.
- Passwords are not to be your name, address, date of birth, username, nickname, or any term that could easily be guessed by someone who is familiar with you.
- Passwords are not to be displayed or concealed within your workspace.

In Rick's summarized version, a password should never be:
- Written, e-mailed, or spoken
- Shared with other people
- Hinted at or made easy to guess
- Used in sync with or duplicated by personal passwords or Web accounts
- Shared when out of the office
- Typed and saved in electronic documents

Despite these good practices, reality will dictate that you will inevitably need to document a password during a project. How can you document a password without compromising general security best practices? In this chapter we will show you how to document passwords and access codes without compromising the confidence you are entrusted with by your employer and customers. You will also be able to document the changes made in passwords over time. Also, in the event that you misplace your documentation binder, you can be confident that it will take serious effort to figure out the passwords that you've documented in your worksheets.

The sample code on the following page contains codes that are both easy to draw and reproduce on a computer (using standard ASCII characters) with a word processor or other text writer. You may use this code or devise your own; the main point being that the code you use should be capable of being reproduced electronically. Characters that are not shown on the sample code such as commas, periods, dashes, etc. can be coded with other characters or just simply used within your codes, being identified with an underscore (____) character. For example, "This is a test for Mon., Tues., and Wed." can be coded "This is a test for Mon_._, Tues_._, and Wed_._," in your respective code or by using substitute codes for these common symbols. These symbols or characters are seldom used, unless you're writing sentences or unique passwords that utilize these symbols. In the latter case you'll want to use special characters as codes instead of using designations to make your coded system more difficult to interpret.

Sample Code Section Table:

Alphanumeric Character	Sample Code
0	ϒ
1	'
2	≤
3	/
4	∞
5	f
6	Ω
7	Ψ
8	Ξ
9	α
10	↔
A	←
B	
C	→
D	↓
E	∘
F	±
G	"
H	≥
I	×
J	Π
K	⇔
L	•
M	÷
N	≠
O	≡
P	≈
Q	…
R	⏐
S	—
T	↵
U	⊗
V	⊕
W	∅
X	∉
Y	⊂
Z	⊇

You will definitely want to standardize your codes if you are sharing your documentation with other technicians, to avoid being asked to decode your security information. If you are having a hard time coming up with a code system, use the sample one that is provided above—just be sure to change the specific associations between character and code.

Use of a coded system

Use the code system exactly as you would use your normal method of writing. One option would be to write right to left instead of left to right. Memorize your code and practice it until you can easily decode without thinking. This will take some time to master, but the result will be knowing your documentation will take some serious effort to be decoded and may require a decode program to transcribe. Whatever you do, *do not* include the code table or reference to the alphanumeric info within your documentation binder or electronic files. If you do, make sure to put a password on the document that contains your reference table.

Here's a sample use of the code system:
NT Server account and password
Acct—admin
Password—2ez2brk4u

Coded documentation
NT Server
Acct: $\leftarrow \downarrow \div x \neq$
Password: $\leq ^\circ \supseteq \leq \ | \ \Leftrightarrow \infty \otimes$

If you are using a software program to track your security attributes it would be a good idea to utilize a 128 bit or better encryption system that can decode on the fly. There are a lot of programs out there that can do this—use the best one that will work with your system.

Here are several steps you can follow to secure documentation containing sensitive information or accounts and passwords:

Step 1: Follow your organization's security policies and procedures. A security policy is a formal statement that specifies the rules all parties must adhere to in order to maintain consistent access to sensitive information. If your policy is not

to document passwords, this will supercede or override the encoding or encryption of passwords presented in this chapter. Our goal is not to make you liable or culpable for violating policy and as a result getting you fired from your job, so whatever your company polices are regarding security (or, for that matter, anything else that the *NuneX Method* may conflict with) supersedes the NuneX standard. Just keep in mind that the *NuneX Method* can serve as a documentation standard where there is none available.

Step 2: Keep up with all security patches and issues as provided by system vendors and software companies. Monitor web sites that expose security breaches, consult your vendors regarding the applicable patches to your network, and then use the *NuneX Method* to plan and implement the patches.

Step 3: Test your security by hiring professional companies to attempt to break in. Test security from a hacker's point of view. Invite a trusted network friend or service that specializes in security analysis to attempt to hack your system. Analyze and follow their recommendations if they are justifiable. Be sure to obtain an initial access agreement from them assuring you they will not reveal their findings or publicize them in any way. Require them to provide you (in writing) their documentation and notes regarding the process and their findings.

Step 4: Your network is only as secure as your weakest link, so endeavor to expose the vulnerabilities of your network. Don't take it personally if you believe your network is secure but you or someone else uncovers weaknesses.

Step 5: Most security breaches occur within an organization, so the best approach is to work from the inside out. Secure all systems internally. Have other system administrators, and maybe some power users, try to access unauthorized areas of the network, and document their efforts. If people with lesser technical skills than you can hack into the network, it is a certainty that curious and possibly disgruntled employees will seek ways to bring the company's network down or reveal sensitive information which may affect your job. Intrusion detection and packet filtering technologies can be utilized for this purpose.

There are many methodologies, systems, and applications that can serve as a guide to security. The main emphasis in Step 3 of the *NuneX Method* is to focus your attention on security and coding your logins and passwords, so that they are not easily revealed within your documentation.

Summary

Step 3 is security and passwords. With the necessity to work on systems and networks comes the need for authorized access. Unless you are the one determining what the passwords are to be in advance, you have to uncompromisingly document the system passwords you require in order to work on the equipment. The need for system access may arise as a result of support to a customer, an internal support call, or backups or system changes. Using a special coding method to securely document passwords for referencing system access is the major part of Step 3. If your documentation falls in to the wrong hands, you will want to protect your interests and deter hacks from the systems you are held accountable for.

Exercises (Answers and examples are located on N-Corp Services' web-site, http://www.n-corpservices.com, under the downloads section):
Referring to the project information below, develop **security codes** for the following projects. Note that some information has been added to each exercise to cover the concepts presented in this chapter. Use the *Security* worksheet on the next page to write down your answers.

Security Worksheet

Page_____   Project _____

Documenter(s): _____

_____

Source: _____

_____

System	Account Name	Password

1. Project—Personal Computer memory hardware upgrade for 50 desktops. Using the coded system worksheet, document the power-on passwords for each system.

Background: You work for a service company as a Computer Technician. You handle break/fix types of service calls, and you've been assigned a small project for an important account, the Smith Company. The project is to perform a personal computer memory hardware upgrade for 50 desktop systems, taking them from 128 Megabytes to 512 Megabytes of RAM. The manufacturer of the systems is Company X and all the desktops are the same model and have the same operating system. You must complete this within three days and your window of opportunity to work on these systems includes early morning, lunchtime, and after hours. The Smith Company's business hours are 8:00 A.M. to 6:00 P.M. They require that the project is finished before their end-of-month processing begins.

Several management PCs will require the use of power-on passwords. There are ten PCs that require power-on or BIOS passwords. The customer's password convention is to use the asset tag number plus the manager's phone extension as a password. Here's the list you have been provided with by your customer contact, Mr. Jones. Using the information provided and your own coded symbols, document the passwords for the ten PCs.

Management Extensions	Asset Tag
1234	S4567
1245	S4468
1256	S4469
1267	S4470
1278	S4471
1289	S4472
1290	S4473
1201	S4474
1212	S4475
1223	S4476

2. Project—Installation of a WIC module for a Cisco 3662 router. Using the *Security* worksheet, document your standard passwords and the passwords for Clinic A via the coded system.

Background: You work as a Network Administrator for a hospital that requires 24/7 up-time from its computer equipment. You've been assigned to a project to add another clinic connection, Clinic A, to the existing main router. The T-1 circuits are now in place. The main router is a Cisco 3662 that already has the necessary memory, firmware version, and slots to accommodate the addition. The only thing you need to do is install a T-1 WIC (WAN interface card) module and at a later time you will program the router and connect it to the previously-configured Clinic A router. You will need to power down the router, install the WIC module, and turn it back on to perform the module insertion. Clinic A is due to be connected and ready to operate twenty days from the day you've been given this assignment. You are now on day fourteen. Clinic A has a PC/Network technician that will be supporting the site and will require access to the router for troubleshooting purposes. The router will require three passwords: a "telnet," an "enable," and a "secret" password. You don't want to compromise your standard administrator password and inadvertently allow the local technician access to your WAN, so you will assign the router different passwords and the "secret" password will be the one you'll use for all the router configurations.

Your standard router passwords are:
Telnet: sys2router_acc
Enable: router_at_site
Secret: _secrets_are_never_revealed!

Your router passwords for Clinic A will be:
Telnet: clinicA_acc
Enable: router_at_clinicA
Secret: _secrets_are_never_revealed!

3. Project—Network infrastructure upgrade for Customer X. Develop a password scheme for your customer and document it using the coded system.

Background: You are a Network Engineer working for a services company that specializes in network design, implementation, and support. You've been assigned to a long-term project for Customer X that encompasses the design and planning of a network infrastructure upgrade from their existing 100-Megabit backbone to 10 Gigabit Ethernet. The design will involve the purchase of new equipment and re-termination of their existing fiber optic backbone to handle Gigabit specifications. You have obtained permission to develop a standardized password system for your client's network equipment. Using your knowledge of security and passwords as well as your creativity, develop a password scheme for your customer. **Note:** You can only use characters that can be typed on a computer keyboard. Your password scheme should include telnet, enable, and secret passwords. You are estimating anywhere from 50 to 100 devices for this design.

4. Project—Remote site password documentation. Document the information provided to you by the Network Administrator for each remote site that you support. Use the *NuneX Method* coded system.

Background: You are a Network Technician responsible for supporting remote sites within a 200-mile radius of the city where your company is located. You have responsibility for three sites: the Finance and Accounting site, the Education and Training site, and the Marketing Department's remote office. Site 1 (Finance and Accounting) is located in City X, 45 miles from your corporate office. Site 2 (Education and Training) is located within your corporate office's main city—except it's in the far west side of town, about 25 miles from your corporate office. Site 3 (Marketing) is located near the corporate office, across the street in a business plaza. Each site has a LAN that connects to the corporate WAN with frame relay circuits ranging from 512K to a full T-1. You are only responsible for the LAN aspects of each site—the WAN support is done at your MIS Data Center by the Network Administrator. The most important site that you maintain is Finance and Accounting since the CFO's office is located there, and its business functionality is vital to the success of your company. Your work hours are 8:00 A.M. to 5:00 P.M. The Network Administrator has given you passwords for the three sites you support for servers, switches, and routers. Document the passwords using the NuneX method coded system. The sites are as follows:

Site 1 Microsoft Windows 2003 Server
Account: LocalAdmin
Password: 1admin@#$

Site 1 Switch & Router Passwords
Telnet: R100107
Enable: R100101

Site 2 Novell Netware 6.X Server
Account: Admin
Password: 2admin@$^

Site 2 Switch & Router Passwords
Telnet: R100108
Enable: R100102

Site 3 Microsoft Windows 2003 Server
Account: LocalAdmin
Password: 2admin@#$

Site 3 Red Hat Linux Server
Account: LocalAdmin
Password: 3admin@#$

Site 3 Switch & Router Passwords
Telnet: R100109
Enable: R100102

5. Project—Develop coded passwords for the systems you are responsible for.
Use the NuneX *Security Documentation* worksheets. Create a word processor document with the same information you enter on your security worksheet. Secure your word processor document by placing two separate passwords—one for modification, and one for viewing.

CHAPTER 4

STEP 4—Notes & Reference Information

The fourth step of the *NuneX Method* describes the process of establishing project notes and reference materials, organizing and saving these, and annotating your data as new information becomes available. The goal of effective project notes is establishing a valuable reference source. Examples of the types of information you'll save to your project notes include sections of an article, a book, magazine, diagrams, or even on-line documentation found in many technical libraries and vendors' archives. The main difference between Step 4 and any other step is that this is information that you didn't *personally* create. In other words, someone else authored the material you'll use in your project notes. Anything you *personally* plan, create, observe, or write should go in the Planning, Security, Documentation, Refinement, or Maintenance sections of your project documentation. Step 4 covers reference materials not produced by you, but rather by another source.

It will be a rare case indeed if you don't have a need to take down notes. However, a situation may arise when you'll simply refer to the reference documentation of previous projects, since service work usually builds on itself. It has been my personal experience that product manufacturers and elite service providers/vendors will usually have better documentation than that which can be produced in-house by most organizations.

How much is too much reference material for your notation section? Don't go to the extent of printing a manual, for more than likely you will not read it all. Try to get the main information you'll need to carry out your project or assignment. If you need to use DOS commands, there is no need to print out the complete DOS command reference with explanations if all you'll need to use is the XCOPY and COPY commands, for example.

Try to keep your notes simple to establish a quick and accurate reference. If you must take more information down, make sure that you read the entire reference and highlight the areas you need to refer to in order to make your reference easier to follow.

There is a saying that a picture is worth a thousand words. In IT, a picture or diagram is an indispensable reference, as it can explain the layout of a network, the logical interconnectivity of a system, and even complex connectivity links or interfaces in a simple way. Use diagrams, charts, and graphs to visually replace volumes of detailed technical narratives.

Resources for reference information can be found in technical journals and publications related to the project or assignment you are working on. Vendor documents and any in-house documentation (if it exists) will be a great resource for researching your project. Ask your supervisor and fellow IT staff in your department for project background details. They can be excellent sources of information from a historical point of view, and can provide insight into why things are being requested or done in a certain way.

What if network diagrams, charts, or manuals aren't available or don't exist for your project? If this is the situation, you must create the reference and make your best effort to document what is within your capabilities to understand and research. For example, you're assigned several different customer sites, and each one of them lacks network documentation. Your main task will be to document as much as you can about each site as you survey them. If you aren't familiar with the products this customer has in place, research as fully as possible each product in order to become thoroughly familiar with them. The documentation efforts in this situation would produce materials you'd include in your notation and documentation sections.

I recall working as a Network Engineer for a medium-sized organization whose network technicians—unfortunately—didn't establish and maintain good documentation about their customers. Technicians were depended on to figure everything out while on-site. When a customer called with a problem, the only way to resolve issues was to dispatch a technician to the jobsite to resolve what would typically be a simple issue. In order for the service work to be effective, someone had to be dispatched to the site. As I made the rounds to each customer I would document his or her network and note important information such as IP addresses, sub-nets, OS versions, system changes, key dates of new equipment installations, etc. I also created a logical diagram of their networks. On each site visit I would update my diagrams and notes, and when I felt I had a pretty good layout I reviewed it with the contact at the site. To the customers' amazement, my documentation proved superior to what others previously had produced, because mine included updated references and timelines. Often I was asked to provide copies of my documentation to the customer for their internal use. Although my company had no official documentation policies, I realized that one site visit wouldn't be sufficient to get the most accurate information, and consequently devoted many hours to this task. The efforts invested in documenting our customer's network saved our company time and money—we didn't need to dispatch technicians to a site for every call, because we were able to resolve most issues remotely. I also researched the products the customer was using that I wasn't familiar with, like third-party switches or wireless products that weren't market leaders. By having some idea of what these devices were, I was able to troubleshoot certain issues related to the equipment that I probably would never have attempted otherwise.

Another good example of notation is what one of my fellow IT network engineers does for his customers. His goal is to be able to rebuild at least 80 percent (preferably 100 percent) of a network for his customers in spite of their MADs (moves, adds, and deletions), over which he has no control. Every time he visits a customer's site he updates his documentation notes and diagrams, and keeps revisions handy to track the changes made to the network. His earnest efforts in this regard have often saved customers in dire straits when network problems arose. Using his diagrams, he was able to track the changes made over time and illustrate to the customers what led to the problems they were experiencing. It takes extra effort to achieve this level of performance—and as a result of these efforts, he's achieved success as a network engineer and now manages other engineers, which means he travels much less. His 20 percent effort in documentation produced 80 percent less travel, allowing him to spend more time at home with his family and earn a promotion. When I interviewed him we discovered he was unknowingly employing most of the steps of the *NuneX Method*. He also credited a major portion of his professional success to having made the extra effort to produce good customer documentation, notation, and reference material to use in resolving support issues.

Verify information

It is imperative that any information you use for reference in your notation section is verified for accuracy and authenticity. There is a lot of misinformation out there that can cause more harm than good. For example, while troubleshooting a virus issue I visited the web-site of the software company whose anti-viral software I was using. On their web-site the company had posted a process to update the virus software and remove the virus. The information didn't seem correct to me, so I called their tech support line. It was about an hour's wait before I spoke to a live person, and when I did speak to their tech support person I discovered the web-site information was erroneous. If I'd followed their posted instructions it would have led to severe system problems. I asked why they'd posted wrong information, and the response was they "hadn't gotten around to updating their site with the correct information," and that the updated solution wouldn't be posted until the next day. Many people calling their tech support line found this out the hard way—they followed the erroneous instructions only to find themselves facing difficult issues. Verify all information in a situation like this (preferably with at least two different sources, if possible) before accepting it as true and useable. If the advice of one technical resource doesn't make sense to you or seems questionable, then obtain a second opinion from another resource—even if it's within the same company or organization.

Organizing information

The final objective is to organize your material for easy reference. If you are using e-mail and faxing as a means of exchanging information or communicating documentation, establish a separate e-mail account or folder dedicated strictly to your documentation efforts. Don't mix personal messages with work or project-related information. The benefit of notation is establishing a guide or train of thought behind your activities, so you (or someone else) can better justify and logically follow the reasoning behind your actions via proper attention to this important step.

SUMMARY

Step 4 is the process of taking down notes and vital information. Notes can be written information, downloaded documents, system diagrams, e-mail messages, memos, notes taken from conversations, excerpts from publications, etc. that you use as a reference or for guidance during your project. The materials that belong in the notation section are those that were produced by someone other than yourself or your project team. The best approach is to collect the information, verify it, and then organize it for future reference.

Exercises (Answers and examples are located on N-Corp Services' web-site, http://www.n-corpservices.com, under the downloads section):

Create notes for the following projects. Please note that some information has been added to each exercise to cover the concepts presented in this chapter.

1. Project—Installation of a WIC module for a Cisco 3662 router. Using various resources, find the commands to set the passwords. Locate, verify, print, and organize the reference information for setting passwords on a Cisco 3662 router running 12.X IOS software.

Background: You work as a Network Administrator for a hospital that requires 24 X 7 up-time from its computer equipment. You've been assigned a project to add another clinic connection, Clinic A, to the existing main router. The T-1 circuits are now in place. The main router is a Cisco 3662 that already has the necessary memory, firmware version, and slots to accommodate the addition. The only thing you need to do is install a T-1 WIC (WAN interface card) module and at a later time you'll program the router and connect it to the previously-configured Clinic A router. You will need to power-down the router, install the WIC module, and turn it back on to perform the module insertion. Clinic A is due to be connected and ready to operate twenty days from the day you've been given this assignment—you are now on day seventeen. Clinic A has a PC/Network technician that will be supporting the site and who will require access to the router for troubleshooting purposes. The router will require three passwords, a "telnet," "enable," and "secret" password. You don't want to compromise your standard administrator password and inadvertently allow the local technician access to your WAN, so you will assign the router different passwords. The "secret" password will be the standard one you'll use for all the router configurations. You don't exactly remember the commands to set passwords on the Cisco IOS software (version 12.X) running on the router. Your task is to research the command set to determine how to set passwords on this device.

2. Project—Network Infrastructure Upgrade for Customer X. Using various resources, find information or references that will verify whether 62.5 μ will support Gigabit speeds.

Background: You are a Network Engineer working for a services company that specializes in network design, support, and implementations. You've been assigned to a long-term project for Customer X that encompasses the design and planning of a network infrastructure upgrade from an existing 100-Megabit backbone to 10 Gigabit Ethernet. The design will involve the purchase of new equipment and re-termination of their existing fiber optic backbone to handle Gigabit specifications. Your customer has informed you that they currently have a 62.5μ multimode fiber optic backbone running at 100-Megabit speeds. Your task is to research and determine whether 62.5μ fiber supports Gigabit speeds.

3. Project—Research any information from various resources on a project you are currently working on. Follow the *NuneX Method's* recommendations for locating, filtering, verifying, and organizing the material.

CHAPTER 5

STEP 5—Documenting

There are many different methods for properly documenting a project that involves technical work. To determine which one is the best approach, consider the easiest and most common form available—by far, the written word. Sometimes it makes sense to combine various forms of documentation such as pictures, audio, video, electronic documents, and other advanced alternatives. Despite the many options for documenting your work, you should consistently keep it in the form your employer or customers require. For the sake of simplicity and universal appeal, the *NuneX Method* will utilize, but will not limit itself to, the written word with accompanying visual aides.

The *NuneX Method* involves documenting your work in the most effective and reasonable manner possible. The main goal is to document the detail required so you can review, enhance, and learn from your documentation at a later time. Further, it is vital that you document so that another technician who may follow you on a project (perhaps with the minimum required technical knowledge and training), is able to understand and reproduce your work effort and results.

To simplify this goal, the *NuneX Method* involves the use of various documentation worksheets for its written word. The *NuneX Method* uses an abbreviation system (known as technical shorthand) for speed and brevity. The *Documentation* worksheet bears some similarity to the *Planning* worksheet, except that it has a procedures section with special line spacing. This special spacing is designated to serve as a prompt for each action taken while performing your technical work. It is important that you code each space as you sequentially record what you did, in technical shorthand. You can also use the special spacing as a reference point to discern whether you are working on the OS (Operating System), NOS (Network Operating System), SYS (System), HW (Hardware), SW (Software), or NET (Network). These are just a few of the prompts you can use as points of reference. On the bottom right of the worksheet is a space for logically representing the work you just completed in the form of a drawing. The grid serves to help you quickly and easily draw what you are actually doing. Remember that a picture is worth a thousand words. You can note "DRW1" (for example) in this space to refer to "drawing number one."

DOCUMENTATION WORKSHEET

DATE: ___/___/200__ DAY _____ TIME: ____:____ AM PM PG ___ of ____ PGS

Project Name:

Documenter(s):

PROCEDURES   Defined Variables:

01____ _____ 31____ _____ 51____ _____
02____ _____ 32____ _____ 52____ _____
03____ _____ 33____ _____ 53____ _____
04____ _____ 34____ _____ 54____ _____
05____ _____ 35____ _____ 55____ _____
06____ _____ 36____ _____ 56____ _____
07____ _____ 37____ _____ 57____ _____
08____ _____ 38____ _____ 58____ _____
09____ _____ 39____ _____ 59____ _____
10____ _____ 40____ _____ 60____ _____
11____ _____ 41____ _____ 61____ _____
12____ _____ 42____ _____ 62____ _____
13____ _____ 43____ _____ 63____ _____
14____ _____ 44____ _____ 64____ _____
15____ _____ 45____ _____ 65____ _____
16____ _____ 46____ _____ 66____ _____
17____ _____ 47____ _____ 67____ _____
18____ _____ 48____ _____ 68____ _____
19____ _____ 49____ _____ 69____ _____
20____ _____ 50____ _____ 70____ _____
21____ _____
22____ _____
23____ _____
24____ _____
25____ _____
26____ _____
27____ _____
28____ _____
29____ _____
30____ _____

There are several things you need to keep in mind as you document on the provided worksheets. You need to be as accurate, complete, and concise as possible. Your documentation needs to be usable both to yourself and those who may need to refer to your documentation later. A logical progression of events needs to be written down. After the project is over and all is said and done, you should rewrite or rework your documentation to create a living, breathing document. This document is transcribed from your technical shorthand to proper English—noting the important sequences and procedures which produced the desired results. You want the documentation to be appropriate to the content and scope of the project. You could also note "things to avoid" for future reference. You'll want your final document to have a high level of quality in all content and information. This process is a major portion of the *NuneX Method's* Refinement phase (Step 6).

You may also want to include meeting minutes or notes you have taken down during project status meetings, phone conversations, and customer site visits. For this purpose, use the *Notes* worksheet. Keep these worksheets handy for quick use and reference. Sometimes information is only verbally dispensed, and having these notes can save a lot of time wasted on trying to figure out what is going on. Providing the parties involved with a copy of meeting minutes can avoid the "he said, they said" problems that occur later when there is no written communication to hold either party accountable. Here is a template that may be utilized to record meeting minutes, followed by the *Technical Meeting Attendees* worksheet and the *Meeting Minutes* worksheet:

Note Taking Worksheet

DATE: ___/___/200__ DAY _____ TIME: ___:___ AM PM PG ___ of ___ PGS
Project Name:

NOTES

_____
_____
_____
_____
_____
_____
_____
_____
_____

Information that is received	**Information that is dispensed**
_____	_____
_____	_____
_____	_____
_____	_____
_____	_____
_____	_____
_____	_____
_____	_____
_____	_____
_____	_____

Diagrams / Charts / Drawings											

TECHNICAL MEETING ATTENDEES WORKSHEET

MTG Date:_____/_____/_____ Day:_____

Meeting Location:_____

MTG Facilitator(s):_____

Start Time: _____ _ AM _ PM   End Time: _____ _ AM _ PM

Reference to MTG Planning Worksheet #: _____

Name	Company	Title / Position	Signature

MEETING MINUTES WORKSHEET

MTG Date:_____/_____/_____ Day:_____

Meeting Location:_____ MTG Recorder/Secretary: _____

Start Time: _____ _ AM _ PM   **End Time:** _____ _ AM _ PM

Reference to MTG Agenda Worksheet #: _____

_____   _____
MTG Recorder/Secretary's Signature   Date

MTG Rules __ New Set __ Addendum __ Revision

Agenda Item / Presenter	Discussion / Conclusion	Recommendation / Action	Follow-up/Responsible Party
Administrative			
Status Updates			
MTG Improvements			
Conclusions			

Attendees List Sheet #: _____
Next Meeting Time: _____ __ AM __ PM
Location: _____

Technical shorthand—the beginner's guide

As mentioned many times before, the IT world is a fast-paced, hurry-up-and-go environment. Due to this nature of the business, IT professionals may find themselves cutting corners and taking the easiest paths to complete projects, meet deadlines, and complete repairs. This can be attributed to the shortages of IT professionals in the United States and other nations in the early 21st century, or even attributed to the budget cutbacks organizations make during financially difficult times. There is too much work and too little time to complete it all. If you have a life outside of work, you may feel that the demands of giving your full attention to your family are a higher priority than taking work home. All these factors contribute to the lack of good documentation in the IT workplace because professionals are usually too busy to write things down or to write them down using a systematic approach. This often explains why some IT shops hire technical writers to do this for them. However, the collected data is usually not clear for these writers, and it often calls for research and further explanation by the originator. A project manager may also contribute to the documentation effort by providing detailed status reports, tracking time and expenses associated with a project. A project manager may provide project closeout documentation in the form of an executive summary and resource utilization report, and any other pertinent information. Despite these efforts, it may be difficult for the customer or project manager to recreate the steps that led to the completion of the project, because the available documentation may lack the necessary detail. There may be no knowledge transfer besides the experience the customer gained as a result of participating in the overall project. The *NuneX Method* covers all these bases (and more) via its use of proactive planning guides that include contingencies, documentation processes for all aspects of a project, provision for knowledge transfer, and provision of summarized information for briefings, and its allowance for maintenance—all designed toward an efficiently and successfully executed project.

In my past personal experiences working in IT, time constraints were usually the reason I seldom documented my work. The fact that I successfully completed my assignments and documented important information during and after a project was good enough for my supervisors. Documentation was an afterthought and something to justify the expenditure of various resources used to resolve an issue or complete a project. Then one day, as the song goes, "I saw the light." Here's one experience that led to my discovery of technical shorthand:

I was preparing to go on an important training trip involving our newly-purchased backup system. The training was scheduled for December 1999—just before the infamous Y2K dilemma. Mine was the last class available for the year, and it was of utmost importance that I attended so that I could help ensure that our enterprise-backup system was 100 percent ready. The vendor who sold us the product implemented it in late November/early December, and as part of the deal we would be trained to properly ensure the product's smooth operation. I was scheduled to fly out on a Sunday morning to attend the training.

At the same time, I was staging a new server for one of our remote sites that was preparing to open doors that following Monday. I was almost done with the setup, except I was encountering a few problems with the custom RAID configuration I was using (we'd purchased a newer RAID controller for the server that I wasn't too familiar with). Before I knew it, the workweek was over and I still hadn't resolved all the issues with the server, and it wasn't ready. I had just one day left to get it going.

During that week I spent a lot of time assisting other techs and analysts with figuring out other problems with other systems, and ultimately cut myself short of time to get the server completed. The continuous interruptions did not let me focus on the project and see it to its resolution. What was planned to be done later had become due *now!*

On Saturday I went to the remote site and began to work furiously on the server, hoping to complete it quickly so I could spend some time with my wife before leaving for the three-day training class. I had the internal pressure of my wife giving me personal grief about working on a Saturday, the external pressure of having to fix the file sever that day, and the added pressure of getting ready to leave the next morning. I realized it would take about a four to eight hour investment to complete the project, test it, and leave notes for my fellow techs to follow what I did in case they needed to troubleshoot the system while I was away.

Instead of writing notes at the end of the project, I decided to write everything down as I worked. This documentation included all error messages, every step, every reboot—I just went crazy, writing *everything* down. I figured that I could filter the information later, and just give my peers what info they needed to know before I took off.

When I was about one hour away from completing the project (around 7:00 P.M.), I rebooted the server. To my aggravation, it didn't come back up. Around the same time my wife started paging me to call her at home. The stress regarding both situations almost killed me. I figured I was going to be in the "doghouse" because I would probably need to work the whole night and redo the server. I figured I needed another eight hours to redo all the configurations, install the software, and set up all security permissions. I called my wife and explained the situation. As an understanding partner, she gave me her support and asked me to do my best and take care of the matter. I then decided to take a small break from the problem, and wondered if there was a way to easily fix it. It was a Saturday night; if I called tech support, I probably would be assisted by the usual what I term "third string" or inexperienced technical support representatives (I have typically had bad experiences calling tech support during off hours and holidays). As I thought more about it, I started to review my notes and detailed documentation, and upon going back a few steps, I soon realized what had gone wrong. I quickly went back to the server and rebooted it a different way. Soon after the reboot I had the server back on-line. Whew! What a great relief! I was then able to finish the last part of the install, testing, and documentation and leave at about 8:15 P.M. Ultimately it was my documentation that saved me. The time and energy that I invested in documenting my work in such detail paid off by leaving me free from any rework, and leaving me able to quickly troubleshoot a serious issue. I got home and—to her amazement and joy—spent the remaining time with my wife. It was a small victory for both my professional and personal pursuits.

The next day, on the flight to the city where the training was going to be held, I reflected on the previous day's events and what I'd learned from my experience. If I had documented most of my projects, I would have achieved success a lot faster and much easier than usual. I thought to myself, I wouldn't mind documenting all the time—except it takes too long and it's easy to lose one's patience when writing everything down. In the fast-paced environment that I work in, with continual interruptions and fires to be put out, I needed a quick and effective way to document.

I remembered that during High School a friend of mine had taken a shorthand course, and using the techniques she learned, she was able to take very good notes quickly and easily and be able to later transpose them to proper grammar. She always took excellent notes during class lectures and while reading and studying. If I used a similar method, I imagined I could probably produce better documentation *and* be able to work within time constraints during a project.

Upon my return from the training class, I started to work on a shorthand method for technical documentation. After drafting the first set of abbreviations, I put these to the test and kept refining them as I worked on other problems and projects. Here's a sample list of the various shorthand abbreviations that I used that are case sensitive. ***Please note that this list is not complete and serves only as an example of a technical shorthand guide.***

Shorthand	Meaning	Shorthand	Meaning
2	To	LO	Log Out
4	For	LPT1	Line Printer 1
1st	First	Man	Mandatory
2nd	Second	Mod	Module
3rd	Third	MS	Microsoft Corp.
4th	Fourth	MTG	Meeting
5th	Fifth	N	No
6th	Sixth	Net	Network
Acc	Access	Nm	Name
Acct	Account	NOS	Network Operating System
Activ	Activation	Note	Notation
ActKey	Activation Key	Nov	Novell NetWare
Add	Addition	NRG	Energy
AlRes-	All Results Are Negative	Nbk	Notebook
AlRes+	All Results Are Positive	OC	On-Call
AV	Anti-Virus	OR	Order
Auth	Authentication	OSR	Operating System Release
Bcode	Bar Code	PC	Personal Computer
Bkup	Backup	PcNm	Personal Computer Name
Brwz	Browse	Persis	Persistent
Cbl	Cable or Cabling	Port	Port
c+a+d	Ctrl + Alt + Del Keys	Prblm	Problem
CSCO	Cisco Systems	Prgrm	Program
CLI	Command Line Interface	Pri	Primary
COM	Communication Port	Prn	Printer
Conn	Connection	Proc	Process
Cmd	Command	Profi	Profile
Config	Configure	Pwr	Power
CPQ	Compaq	Qry	Query
DAT	Data	Qty	Quantity
DEC	Digital Equipment Corp.	Quote	Quotation
Discon	Disconnected	Rdr	Reader
Def	Default	Rbt	Reboot
DELL	Dell Corp	Rec	Received
Demo	Demonstration	Res	Results
Dev	Device	Resrz	Resource
DnGrd	Downgrade	Roam	Roaming
Dom	Domain	Rpt	Report

Dsktp	Desktop	Rtr	Router
Dwnld	Download	Rx	Pharmacy
Eth	Ethernet	SCSI	Small Computer Systems Interface
Elec	Electrical	SN	Serial Number
ENG	Engineering	Sub	Subnet
Entrprz	Enterprise	Srvr	Server
Env	Environment	ST	Start Time
ERR	Error	Stnd	Standard
Escal	Escalation	-	Subtraction
ET	End Time	Sw	Switch
EQ	Equipment	SW	Software
Fil	File	Set	Setup
FC	Flow Control	SurgeP	Surge Protector
FW	Firmware	Sup	Support
GW	Gateway	Tech	Technician
Hub	Hub	Telnt	Telnet
Hdsk	Help Desk	Tstd	Tested
Hom	Home	TelCo	Telephone Company
HC	Hot Chassis	Ttl	Total
HW	Hardware	UpGrd	Upgrade
IBM	IBM Corp.	Usr	User
IMG	Imaging	UsrNm	User Name
IPA	IP Address	Ver	Version
Inet	Internet	Vir	Virus
IRQ	Interrupt Request Line	VLAN	Virtual Lan
Key	Key	VT	Video Terminal
LAB	Laboratory	W2K	Windows 2000
Lbl	Label	W95	Windows 95
LgOn	Log On	W98	Windows 98
LI	Log In	WIN	Windows
Lic	License	Wkly	Weekly
Lnk	Link	Wknd	Weekend
CL	Closed off	WNT	Windows NT
OP	Open	Xmit	Transmission
Off	Office	Xover	Cross Over
Tlbr	Toolbar	XP	Windows XP
WME	Windows ME	Y	Yes
WXP	Windows XP	Y2K	Year 2000
Run	Run Program	Start	Start Program

What is shorthand? Shorthand is a system or form of rapid handwriting employing symbols and abbreviations to represent words, phrases, and letters for reference. Let me set some realistic expectations: technical shorthand is not learned overnight by reading over abbreviations and then quickly utilizing them. Even though some abbreviations are self-descriptive, it takes practice to become thoroughly familiar with them and achieve consistency.

Here's an example of how technical shorthand can be used to save time.

This is the long way of documenting a program installation on a Windows XP system:

Operating System: Windows XP Professional
Technician: Richard Nunez
Start Time: 2:00 P.M.
1. Closed all open programs
2. Closed anti-virus software
3. Closed MS Outlook 2003
4. Clicked on Start
5. Clicked on Run
6. Typed in "D:\setup.exe" to begin installation program
7. Followed all default settings
8. Rebooted PC
9. Tested Operation
10. Installation Successful
End Time: 2:30 P.M.

Here's the technical shorthand version:
OS: WXP Pro
Tech: RNunez
ST 200P (**Note:** You may also use military time designations, such as 1400 hrs.)
1. CL OP Prgms
2. CL AV
3. CL MS Outlook
4. Start
5. Run
6. Set
7. Def
8. Rbt
9. Tstd
10. AlRes+
ET 215P

For ten steps, writing the long way doesn't seem too bad; however, imagine a three-hour detail-by-detail installation process. It would be more preferable to use the shorthand considering time constraints—especially if you have to write it by hand.

When you think about shorthand, it is nothing really new to computer professionals, especially those who have done any programming or coding. Programming is the use of shorthand commands to carry out complex instruction sets. Think of shorthand documentation as programming and it may become easier to learn and adopt as a new method of documenting your work.

The best way to learn technical shorthand is to either use the provided glossary of abbreviations, or to develop your own shorthand. I recommend that you use whatever makes it easier for you if you're documenting on your own. If you adopt a method for a department or company, it would be prudent to adopt a standard so that everyone's documentation is in the same format. The universal shorthand abbreviations provided on the table shown earlier can be used as a basis. Updated abbreviations will be provided on our website at: http://www.n-corpservices.com. In future website postings, we hope to develop a universal database and learning tool that will be readily available to those who wish to standardize their documentation by using technical shorthand symbols provided by N-Corp Services for the *NuneX Method*.

What is the best format for documenting in technical shorthand? In IT, there is usually a need to document procedural steps, time, notes, names, information, and diagrams. Use the format that makes it intuitive for you and your team to later transpose. Time invested in detail can pay off later when things go wrong or go unexpectedly—which, unfortunately, is the way most projects go.

Documenting support calls

Be sure to use the *Notes Documentation* worksheet for step-by-step procedures to record information a tech support person told you. Note the full name of the person you spoke with, his or her phone contact information, his or her supervisor's information (if necessary), the date and time you called, the results of your call, and any action items that were promised by either party. Use the worksheet or—if you are a poor note-taker and your phone system has the capacity to do it—use the voice recorder feature to record your conversations (assuming it is legal in the state you live and work in) and later play it back to transpose to written documentation. Some systems may provide the capability to export to an audio file format and to electronically attach or store the conversation for future reference.

Documenting potential issues/concerns

What do you do when during the course of the project you discover potential issues that may severely affect its outcome and are beyond your control? You need to document these concerns and share them with the proper authorities. Send a written memo and follow it with a phone call and meeting request, or vice versa. The goal is to let your concerns be known so that no blame may come back to you if the situation worsens. The power of a formalized document is very effective in getting results from people, especially in a bureaucratic environment.

Documentation levels

Document everything you can think of. The importance and priority level of the project, since it has an impact on your business, determines the level of detail and effort needed in your documentation. There is a direct relationship between the *importance of* and the *level of detail for* each project. If you are doing a hard drive upgrade on a PC, you may choose not to document at the level of detail you would if you were working on a hard drive upgrade on a critical file server that has been taken off-production during a peak usage time. The size and importance of the job determines the efforts put into the documentation. In essence, use your best judgment regarding how much detail to document and what is necessary.

Documentation Level Varieties

Briefings—These are short, factual pieces of information that give instructions or preparatory information to someone. The documentation includes bullet points, sentence length statements, brief how-to types of instruction, and a list or sequence of occurrences and observations.

Summaries—These are presentations of the substance of a body of material made in a condensed form via reducing it to its main points. A summary includes the same information as a briefing, except it expands to provide slightly more data and information. The summary is written in paragraphs and will usually not be more than a few pages in length.

Details—Details are particulars considered individually as well as in relation to the whole. This level of documentation includes everything that is done: notes, observations, and the project's results. The length and detail of the data provided may cover many pages, containing information that is included in both briefings and summaries.

If you're working short projects that don't involve a lot of detail, using the *Documentation* worksheet may be overkill. Alternatively, you may opt to use the *Short Term Project Documentation* worksheet. The condensed form doesn't necessarily require the use of technical shorthand, but does require answering key questions and noting certain points related to your project.

A standard documentation format to use for performing short-term projects or technical tasks for everyday use follows here:
1.   **Verify the problem or customer request.** Is what your contact or customer describes as the problem *really* the issue? By doing problem determination, did you come upon a different issue? If so, briefly state what the real problem is and any associated error messages to better prepare for the project.
2.   **Note equipment name, model number(s), serial number(s), and contract support number(s) for the associated equipment.** This will be necessary for tech support calls and for references within your documentation.
3A.   **Answer the following questions regarding the Problem or Request:**
Who—Who is involved and who is affected? Who is going to pay the bill? Who is the point of contact (POC)? Provide as much contact information as available—such as addresses and phone numbers.
What—What business processes will be affected if the problem is not resolved, or if the request is not performed? What are the business needs and requirements?
When—What's the absolute deadline for resolving the problem or for performing the requested service work? What are your windows of opportunity to get this done? Indicate the times you're able to work the issue, or what day and time you schedule the response call for.
Where—Where is the facility, building, room, etc. located? In which office, room number, or area is the technical work going to be performed?
Why—Why are you engaged in this service call or request? Why are your services needed?
How—This is addressed via the *Planning* worksheet or briefly state the main idea behind your efforts.
3B.   **Stop and start times, along with dates.** Provide hourly summaries to reference dates and times. Use a date & time stamp log to show historical response and activity.

4. **Answer the following questions regarding the Resolution:**

<u>Who</u>—Who helped you resolve the problem or perform the requested service? Provide the tech support numbers, names, extensions, and anything related to people who assisted you with this project. Who was the point of contact?

<u>What</u>—What did you do that *worked* to resolve the problem or requested service? Also, what did you do that did *not work*? What did you learn of significant importance during this project? How did you do it? Provide feedback on the *Planning* worksheet.

<u>When</u>—How long did it take you to resolve the problem or perform the requested service? Note the elapsed hours, days, weeks, months, etc. Accuracy is vital in documenting timeframes, deadlines, etc.

<u>Where</u>—Did you have to go to various sites to resolve this issue? If so, document the locations where you had to go to perform the service work.

<u>Why</u>—Why did the "what" portion of your resolution work? Why did it *not work*? This is a reflection on your service work. Keep in mind you may or may not have the appropriate answers. In such cases state your theories and ideas.

5. **Observations**—Note any hidden issues that may have affected the performance of your work assignment. Did the customer not have a wall outlet for power? Did the customer not provide you with access to the area you needed to work in, and as a result delay your visit? Document any observation that needs to be noted for potential billing issues or complaints from the customer. Also note anything that may benefit another technician when planning a visit to the customer—such as a need to sign in at the security desk and request an ID badge. Include maps, toll charges to get to the site, landmarks to identify where the site locations are, etc. Also note whether the customer may require training on a particular skill set to avoid the problems he or she created.

6. **Miscellaneous, notes, etc.** This covers anything else that may be pertinent for this call not explicitly covered in the previous six steps.

The following is a worksheet you may use to document these points:

Short-term Project Documentation Worksheet

DATE: ____/____/200__ DAY ____ TIME: ____:____ AM PM	**PG** ____ of ____ **PGS**
Project Name:	Start Date:
Documenter(s):	Completion Date:

Customer Name: _____	Contact Name: _____
Customer Address: _____	Contact Phone: _____
	Contact Cell / Pager: _____
City _____ ZIP _____	Contact e-mail: _____
Work Hours _____ Time Zone _____	After Hours Contact: _____

Ticket or Case #:	Internal	External

Problem Description or Service Request

_____

_____

_____

Steps 1 - 3

____ _____
____ _____
____ _____
____ _____
____ _____
____ _____
____ _____
____ _____
____ _____

Steps 4 - 6

____ _____
____ _____
____ _____
____ _____
____ _____
____ _____
____ _____
____ _____
____ _____
____ _____
____ _____

Documentation Process:		
Step 1: Verify Information	Step 2: Equipment Information	Step 3: Answer H5W for Request
Step 4: Answer H5W for Resolution	Step 5: Observations	Step 6: Miscellaneous

H5W: Who, What, When, Where, Why & How

Completed By (Print Text):	Signature:
Work Acknowledged By:	Signature:

Note-taking

Note-taking skills are based on your lifelong experiences and ability to listen, assimilate the information you were presented with, and then note the main points on paper or by other means. There are many ways to take notes; it is important for you to use what best works for you combined with some simple organization, and to review immediately after your note-taking session is over so that you may transcribe your notes into a workable format for later reference and review. In other words, your notes need to make sense long after you have taken them.

During my early college years, the classes I attended sometimes had hundreds of students per class. It was difficult to get a seat at the front since the early bird students often got the best spots. The professors usually talked into microphones so that everyone attending could hear them. They often drew diagrams on chalkboards or whiteboards as they spoke. It was very difficult to take notes quickly and keep up with the visual aides. What I and many of those who had difficulty hearing did was first get permission from the professor to tape record their lectures, then place a mini-cassette tape recorder in an area where you could clearly record without as much background noise (I labeled my recorder with my name, in case it got misplaced), and record their lectures. I made sure I had enough tape to accommodate the class time.

I would take notes at the same time that I recorded the lecture. Later that evening or the next day, I would put headphones on and play back the lecture. To my surprise, I regularly found out I'd missed at least 20 to 40 percent of vital information after reviewing the tape. This process gave me the edge I needed to obtain better grades, because the professors I had often based their exams on their lectures rather than on the assigned readings. Recording the lectures and later transposing the missing information to my class notes allowed me to better understand and learn the information that was presented.

This process may not be necessary for most work environments. If you are working on a project with subject matter you are not thoroughly familiar with, though, it may be wise to tape or even videotape your activities, meetings, trainings, or any other session you may be a part of to review and ask questions about them later.

Documenting phone conversations

The majority of our communication during a project endeavor occurs via e-mail and voice communication, using standard technological devices such as a regular land line phone, a cellular or wireless digital phone, or a radio. Oftentimes critical information is relayed verbally. How can you ensure that this information is properly documented so that misunderstandings are minimized?

Conversations are best documented by immediately taking down notes on what is being discussed (during the conversation), and then later transposing your notes to a report document, noting important dates, times, people's names, phone numbers, discussed topics, your responses, etc. If the person you are conversing with has e-mail, then send this report to him or her electronically for verification. If they object, ask them to post a reply to you detailing their objections, and then resend your modified document until all parties agree on the content. If they don't have a means of electronic communication, then send them your report in a memo or letter format, also asking them to verify the information and to get back to you to clarify any issues. If you have the capacity to do so and if it is lawful in your state, you may wish to record your conversations and later transpose them to a report. Keep the tapes you record handy for proof of your conversations with your project's contacts.

This takes a lot of extra effort and time to do. However, it will be justified if there are communication issues or if, for some reason, the people you spoke to don't come through with their commitments—or you are being blamed for something beyond your control. Your documentation will save you and force others to take you and your project involvement seriously. Your associates will know that you are a documenter, and that anything that is key and vital to the successes of your projects will not evade your full attention.

There are several worksheets that you may use to document your conversations. The *Communication Log* tracks verbal communication activities. The *Note-Taking* worksheet can help you document the details of your conversations, or at least the main points. You should later transpose the information to a report format and file it in the documentation section of your documentation binder.

Communication Log Worksheet

Project: _____ Page ___ of ___

Name of Company or Individual Called R=Received call from C=made Call to V=Voicemail	Phone Number Dialed P=Phone C=Cell PGR=Pager ~ = Approximately	Date & Time A=AM P=PM	Result B=Busy LVM=Left V-Mail LMwP=Left Message w/Person NR=No Response

Documenting meetings

As I mentioned in the Planning section of the *NuneX Method*, meetings are an important part of dispensing and communicating information. Your preparation for conducting or facilitating a meeting should optimize the delivery and presentation of materials. You should also make arrangements for someone else to take good notes (or record the meeting) for later review and generation of the meeting's minutes. There will be times when you are just an attendee and need to be present for a meeting being conducted by someone else.

As part of your documentation section of your project binder, you will need to include any notes taken in or dispensed by these meetings. You can use the *Notes* worksheet or the *Meeting Minutes* worksheet to document key information presented during a project meeting.

Documenting expenses

If you already have a corporate policy on documenting and submitting expense forms, then those guidelines will supersede the following information. Expenses are not something to take lightly—especially if it's *your* money that is being borrowed by your employers to conduct their business affairs. This may include your own purchase of airfare, rentals, meals, hotel, gasoline, etc. You may be using your own money to finance your employers' business activities and your employers will reimburse you later when the proper paperwork is submitted for their review, verification, approval, and final processing—which can sometimes take from several days to several months. If you are using credit cards to finance your business activities and you get paid weeks later, you may be hit with additional interest costs. The best policy is to get everything submitted for processing as soon as possible and to not rely on someone else to do this for you—even your secretary. I've often made the mistake on relying on others to take care of submitting expense forms and requesting reimbursements for me, with very disappointing and frustrating results. My bad experiences with having others handle my company reimbursements eventually caused me to lose money or even jeopardize my credit standing due to the long periods of waiting for reimbursement. *You* are best suited to handle your own financial transactions, unless you know that the person submitting your expenses will do so in the most accurate and timely manner.

The best thing to do when documenting expenses is request a corporate credit card that your employer will pay directly. Most credit card statements are now available on the Internet and can be used for documentation, along with the receipts that you keep for processing. If this is not possible, then as part of your planning request pre-payment checks for travel or other business activities that may require you to personally initiate the financial transaction. Verify ahead of time that there are no corporate freezes on travel and expenses, and that the expenses you are preparing to pay for will be later reimbursed. Get authorization to do so in writing so that you can later present this along with your paperwork, if necessary. Do everything you can during the planning stage so that you don't have to deal with the headaches associated with reimbursements.

I've seen and heard of people leaving organizations due to their company's gross handling of reimbursements (or the lack thereof). Why should you be penalized for conducting business activities for someone else who's using your own money? Avoid these issues by documenting all receipts, making copies of your submittals, filing these in your documentation binder under the documentation section, and finally by submitting all the appropriate forms to your employer for proper processing as soon as possible. Get advanced approval in writing from senior management so that you don't get questioned later. If you don't receive payment within the allotted timeframes, follow up on these issues immediately—don't wait too long or you may be stuck with making large payments you can't afford to make on your credit card. If your company is going through some rough financial times, more than likely you will not see your reimbursements processed as quickly as you would like.

If your expenses will be billed to a customer later, make sure you have their authorization to bill your expenses ahead of time and make copies of your expense documentation for their records. If the customer is paying for your expenses, then request their recommendations on where to stay and what is reasonable as far as hotel and lodging, meals, rental car, etc. I've experienced issues wherein customers have later questioned the expenses of on-site support personnel and have held up payment accordingly. Get pre-clearance prior to any travel—especially in writing.

Documenting repetitive work

Should you document every time you do repetitive work? The answer depends on whether the project is an everyday task or a specific project that may involve some variance from the norm. For example, if you are the server guy or gal and perform server configurations daily, you may want to only document those that vary from the norm or have special or important significance to your business operation. If you are assigned various server installations that will have the same configuration and setup, you should still have performed the first four steps of the *NuneX*

Method—Pre-Documentation procedures, Planning, Security, and Notation. Having done this preliminary work, you can later use the collection of these documents to quickly carry out other similar projects, thus allowing for the 80/20 rule of 80 percent work being done via 20 percent preparation. You will definitely need to document at least the first couple of setups until you have achieved a perfected method and consistent approach to the remaining servers that you are configuring. It may take anywhere from two to four setups until you feel you have achieved the perfect method of setting up and configuring this group of servers. Once you have achieved the perfect model for the server configuration and setup, then you can develop a flowchart to use later for the Refinement stage, or a checklist for other people to follow.

SUMMARY

Step 5 is working your plan (i.e., the plan you made in Step 2). Follow your plan and implement your contingencies, if necessary. This is where the real documentation and step-by-step accounting of your work efforts takes place. Utilization of documentation tools (such as technical shorthand) will ease the pain of laboriously writing everything down, especially when under time constraints and pressure. Document everything as it happens, taking down notes and observations. Technical notes and references used are also noted here. It is important to document phone conversations and potential issues that you may have discovered that may affect the project. The level of detail in documentation depends on the importance and priority level of the project. It is important that you document your expenses and get pre-approval to minimize any issues related to reimbursements. It is also critical that you don't document in a hurried manner so the quality of your documentation won't be compromised. Repetitive work may need to be documented if the project assignments vary from the everyday norm or have a special affect on your business operations.

Exercises (Answers are located on N-Corp Services' web-site, http://www.n-corpservices.com, under the downloads section):

1. **Project—Personal computer memory hardware upgrade for 50 desktops. Using the *Short-Term Project Documentation* worksheet, document the project as you carry it out—refer to your *Planning* worksheet for procedural information.**
Background: You work for a service company as a Computer Technician. You handle break/fix type of service calls, and you've been assigned a small project for an important account, The Smith Company. The project is to perform a personal

computer memory hardware upgrade for 50 desktop systems, taking them from 128 Megabytes to 512 Megabytes of RAM. The manufacturer of the systems is Company X and all the desktops are the same model and have the same operating system. Your deadline is to complete this within three days and your window of opportunity to work on these systems are early mornings, lunch time, and after hours. Smith Company's business hours are 8:00 A.M. to 6:00 P.M. They require the project to finish before their end of month processing begins.

2. Project—Installation of a WIC module for a Cisco 3662 router. Using the *Short-term Project Documentation* **worksheet, document your project approach.** Background: You work as a Network Administrator for hospital that requires 24 X 7 up-time from its computer equipment. You've been assigned to a project to add another clinic connection, Clinic A, to the existing main router. The T-1 circuits are now in place. The main router is a Cisco 3662 that already has the necessary memory, firmware version, and slots to accommodate the addition. The only thing you need to do is install a T-1 WIC (WAN interface card) module, and at a later time you'll program the router and connect it to the previously-configured Clinic A router. You will need to power-down the router, install the WIC module, and turn it back on to perform the module insertion. Clinic A is due to be connected and ready to operate twenty days from the day you've been assigned to the project. You are now on day nineteen. Clinic A has a PC/Network technician that will be supporting the site, and will require access to the router for troubleshooting purposes.

3. Project—Network Infrastructure Upgrade for Customer X. Refer to your *Planning* **worksheet and network design, and document the project in the most logical progression. Note:** *do not* **use technical shorthand to document this project.** Background: You are a Network Engineer working for a services company that specializes in network design, support, and implementations. You've been assigned to a long-term project for Customer X that encompasses the design and planning of a network infrastructure upgrade from an existing 100-Megabit backbone to 10 Gigabit Ethernet. The design will involve the purchase of new equipment and re-termination of their existing fiber optic backbone to handle Gigabit specifications. The customer has already issued the purchase order and the equipment has already arrived at your configuration site for pre-staging. Everything is set to begin the project, and you will begin the implementation phase during the weekend. The necessary cabling configurations to support 10 Gigabit Ethernet have already been completed. You have a crew of two network engineers and four network technicians assisting you with this project. You are the project lead and the documenter. See the network diagram sample in Chapter 1, and use it as the preferred design model.

4. Project—Remote site outage. Use the *Documentation* worksheet in conjunction with your *Planning* worksheet to document the troubleshooting of the site. Note: Use technical shorthand technique for this exercise.

Background: You are a Network Technician responsible for supporting remote sites within a 200-mile radius of the city in which your company's corporate headquarters is located. You have responsibility for three sites: the Finance and Accounting site, the Education and Training site, and the Marketing Department's remote office. Site 1 (Finance and Accounting) is located in City X, 45 miles from your corporate office. Site 2 (Education and Training) is located within your corporate office's city—except it is in the far west side of town, about 25 miles from your corporate office. Site 3 (Marketing) is located near the corporate office, across the street in a business plaza. Each site has a LAN that connects to the corporate WAN with frame relay circuits ranging from 512K to a full T-1. You are only responsible for the LAN aspects of each site—the WAN support is done at your MIS Data Center by the Network Administrator. The most important site that you maintain is Finance and Accounting, since the CFO's office is located there and its business function is vital to the success of your company. Your work hours are 8:00 AM to 5:00 PM. It's early on a Monday morning (around 7:15 AM), and you receive a page from the MIS Help Desk. It seems the users in the Finance site can't log into the network—or if they do, their connectivity goes up and down. You grab your documentation binder and arrive at the site. You review your plan and start your troubleshooting project.

5. Document a project that you are currently working on or one you've just started. Be sure to use your *Planning* and *Contingency Planning* worksheets as a reference. If you conduct meetings for this project, use the *Meeting Minutes* worksheets to provide project status meeting reports. Document your expenses.

6. Use the *Note* worksheets to document a meeting that you are going to attend. If possible, get permission to record the meeting and later compare the recording with your notes. Based on your comparison, how accurate are your notes?

7. Use the technical shorthand list (provided in this book) to document Project 3 (the Network Infrastructure Upgrade, above).

8. Transpose your technical shorthand documentation from Project 4 (the Remote site outage, above) into proper grammar.

9. Document a recent support call or phone conversation that you had using the *Communication Log* and *Note-taking* worksheets.

CHAPTER 6

STEP 6—Refining Your Documentation

When I think of refining documentation, I think of the process of refining precious metals. Craftsmen work at a refinery, taking raw material and purifying it by super-heating, melting the ore, and finally drawing out the raw silver or gold. In time the raw material will eventually separate—the dross remaining near the top of the refining-pot, and the purified metal retrieved from the bottom. The process transforms raw material of little use or value into something pure and precious. The most valuable bronze, the strongest iron, and the purest silver and gold were created only by being subjected to the intense, melting heat of the refiner's fire.

Like the craftsman who draws out precious metals using an intense fire, the technical documenter must also transform his collection of project information into a valuable end product. This is done by intensely reviewing, organizing, and summarizing one's documentation en route to a usable format.

When you finish documenting, the next step is to clean up your notes so they reflect the successes and the failures of your project. This period of afterthought and reflection will improve your processes and will enable you to become more effective. Refinement is achieved by reviewing your activities and the methods you used to collect information, and then using your project notes to create a formal document for both external and internal readers. *The sooner you perform the refining process, the better.* Your peers, customers, and superiors will refer to your project document when they have a need for the important facts contained in it.

Here are some of the characteristics of good refined documentation:

Characteristics of high-quality documentation
- Accuracy
- No omission of any steps in the process
- Logical progression
- Language the reader will be able to comprehend
- Written so the reader gets it the first time around
- Verify correctness of everything—grammar, spelling, step-by-step process

(Adapted from *The Art of Technical Documentation,* 2nd Edition, by Katherine Haramundanis, Digital Press, Boston, 1998.)

You'll want your refinement process to be a reflection of your best work. Strive to be as accurate as possible. Be sure that important information is not omitted during the process of refining. Following a logical progression so another IT professional will be able to follow and understand your methodology without question. The level of effort you put into an important legal response document should be the same you use for refining documentation. Consider your final project document as formalized communication to the stakeholders of your project, those who have a vested interest in its outcome.

It is important that you apply your best efforts in producing a successful documentation product you can deliver as part of your post-project activities. The main theme throughout the utilization of the *NuneX Method* is the pursuit of success. "Desire is the key to motivation, but it is determination and commitment to an unrelenting pursuit of your goal—a commitment to excellence—that will enable you to attain the success you seek."–Mario Andretti

To refine means "to reduce to a pure state, or purify." It also means "to acquire elegance." You (the writer), as well as the reader, should view the refined document as something of purity and elegance. It is important to note here that you should never write and/or turn in your refined documentation the day it was written. It is preferable to have someone else edit and proofread your work prior to submitting it to the project's stakeholders. If you don't know anyone who can do this, there are professional editing companies who can do it for you. You should also work on this document during some quiet time when you can write undisturbed and be clear, concise, and thorough in your communication. In an article written by David Parkinson, MCSE, CNA, on TechRepublic.com, entitled "Documenting Projects Leaves a Professional Account of Your Work" (May 31, 2000) he mentions some key points regarding documentation for consultants. He noted that there are three concepts that are critical to the Refinement stage of documentation: **traceability, history, and control**. It's important to track configuration changes during the execution of a project and the subsequent maintenance phase. This is a good example of documentation providing traceability, though project documentation can also track the sequence of implementation steps, the key decisions made in the project, and the time scales of the project. The documentation can also provide a history of the project as a constant work in progress, through the addition of daily or weekly updates. Documentation should also provide a sense of control for the project by supplying a single source of knowledge reviewed by all parties and understood as the accepted definition of the project's current state.

Communication format

When communicating the important aspects of a project, use a good standard for presenting your information to your audience. An effective way to do this is to:

1. Know your customer and tailor your approach to him/her/them.
2. Prepare in three stages:
   * Outline your ideas—follow the seven logical steps of the *NuneX Method*.
   * Develop the narrative and refine the document.
   * Design supplemental materials.
3. Outline your report in this order:
   * Project objectives
   * The body (key points, supporting material, transitions)
   * Introductory preview and closing summary
4. Use a type font like Arial or Times New Roman to allow your readers to read your documentation without getting eyestrain.

In essence, you should convincingly explain to your audience what you are about to tell them from an introductory standpoint, then go about telling them the key points, and finally repeat what you just told them in a summarized form. If applicable, provide them a reason (or reasons) to take action based on your recommendations.

Documentation templates

You should write at least one **external** document summary that covers the main points of your project, such as the project scope, deliverables, timelines, key participants, issues, financial information, results, and key observations. You should also write an **internal** document summary that covers the same items as the external document with the addition of the knowledge management aspects of the project (such as what was learned, what worked, what didn't work, bits and pieces of information learned from the project, and so on). Depending on your intended audience for the external document, you may write your internal document as an executive summary or technical briefing. An important goal to keep in mind is the ongoing usability of the document, either for your own future use or that of someone else who may support your project in the maintenance phase. The information should be presented in a format that allows others to understand and follow the project's progress, and be able to quickly get up to speed on what they need to know to support it. A good example of readers of your document is the help desk staff. You may opt to use the *Refinement Document Template for External and Internal Audiences* on the following page.

Refinement Documentation Template—External audience outline:
Report Date, Day, & Time

Name of Report Writer

Introductory Preview

Project Background & Scope (Include the business needs for this project)

Project Objectives & Deliverables

Project's Financial Information (Revenue, expenditures, and return on investment)

Project Timeline

Project Team Participants (with department names, e-mail, dates, and phase
participation)

Project Issues, Transitions, & Key Observations

Project Results

Project Conclusion Statement/Summary

Project Support Information

For internal audience communication add the following:

1. What aspects of the project worked well and achieved desired results (i.e.,
    did the project plan work?)
2. What aspects of the project didn't achieve desired results (i.e., did you need
    to use contingency plans?)
3. What did you personally learn by working on this project?
4. Your personal observations of people, time constraints, resources, equipment,
    software, tools, etc.
5. If you had to work this project all over again, what would you have done
    differently?
6. An overall synopsis or summary of your version of the project—the whole
    truth about it.

Knowledge management

As you learn new things while working on the project, utilize the *Notes* worksheet
to write these down and put them in the documentation section area of your
documentation binder—don't wait until the end of a project to do this, or you'll
forget the details. When you complete the project, go back to reflect and expand
on this information, and include it in your internal refinement reporting.

What is knowledge management?

According to Megan Santosus & John Surmacz (writers for the web-site CIO.com) in an article titled "The ABCs of Knowledge Management": "Unfortunately, there's no universal definition of KM (knowledge management), just as there's no agreement as to what constitutes knowledge in the first place. For this reason, it's best to think of KM in the broadest context. Succinctly put, KM is the process through which organizations generate value from their intellectual and knowledge-based assets. Most often, generating value from such assets involves sharing them among employees, departments and even with other companies in an effort to devise best practices. It's important to note that the definition says nothing about technology; while KM is often facilitated by IT, technology by itself is not KM."

Knowledge management, as it applies to the *NuneX Methodology*, is the documenting of knowledge, observations, training, and experiences gained while performing a project, followed by the placing and organizing of these bits and pieces of information in your Refinement section to be later reviewed and communicated to an internal audience. The internal audience is composed of your fellow team members or any other resources you may wish to share this information with. It's a win-win approach to saving time, energy, training, resources, and money due to constant changes in personnel and resources that provide support to the project (or will to a similar one in the future). You may later collect all internally-communicated refinement documents and sort them, catalog them, and organize them in a database for easy access, research, and retrieval. This collection of refinement documentation allows for the development of a home-grown knowledge management system. The key aspect of this endeavor is keeping the data collection consistent, organized, useful, and meaningful.

Honest reflection

Depending on whether you can trust your supervisors, management, and employer, you may optionally provide truthful details regarding a project's execution. Maybe the reason your project took longer to complete was due to the lack of performance of a key member of the project team. This is where you can document and report these issues. They can later be used for performance counseling sessions or (in some extreme circumstances) disciplinary action. Maybe another team member reviewing your refinement documentation reads the honest truth about a vendor's lack of support and decides to look for new support resources. If there is a lack of honest communication, business mistakes will continue to be made. If a knowledge management system is devised to include honest reflection without penalizing those who contribute, upon review of key information the organization may seek out ways to improve its overall performance. The key point is to not focus on placing blame, but to resolve the issues that will contribute to the overall success of the organization. It takes a mature team of individuals to objectively review this type of information and responsibly act upon it.

Your Documentation Can and Will Be Used Against You

You also need to become aware of potential legal issues and liabilities that can result from honest reflection in published reports. I personally have witnessed system engineers and other technical staff experience lawsuits as a result of being too honest in their external communication reports to key customers. These reports somehow found their way back to competitors who were truly to blame for the initial problems the customer was experiencing. Utilizing the external communication reports that included the honest reflection, the competitor initiated restraining orders and later lawsuits against the engineers who produced the report, stating that they lost business as a result of the engineer's "bad mouthing" of their previous service work.

I recommend the following procedures in order to avoid having your documentation used against you:

1. Have your client/customer/stakeholder sign a non-disclosure agreement to make sure your work does not fall into the wrong hands and get used against you. If it does, you can hold the entities that revealed it to be liable. Make sure your non-disclosure document is legal in the country and state you live and work in.

2. Never send out a document without first proofreading it to ensure it wasn't written in anger or an emotional state with the objective of causing damage to anyone. Having a waiting period and asking an objective reader to review your documents prior to sending them out is a good safety net. Be ready to take advice and put yourself in all stakeholders' shoes—even your competitor's. Remember, people will generally take action against those they feel have caused them undue harm.

3. Never blame or outright accuse anyone or any company, entity, or organization of doing wrong, directly placing blame, etc. in your communication reports. Let the reader come to his or her own conclusions via justification, proof, and logical reasoning. Stick to the provable facts and figures. Let the decision-making be done by the reader.

4. When in doubt, have a legal professional such as an attorney review your reports in order to minimize your liabilities and exposure to lawsuits.

By following best practice and common sense, you'll have great documentation without the pains of it having come back to haunt you at a later time.

Using your documentation to further your career

If your employer or customers will allow it, you may request their permission to use your refinement reports to keep a personal portfolio for later use in obtaining new assignments, projects, or career advances. Meredith Little (in an article written for TechRepublic.Com, posted November 19, 2001, entitled "Including Clients Projects on Your Portfolio") wrote some great material on how to do this. You may want to include within your standard engagement contract a clause stating that you have the right to retain one copy of all documentation you create. This clause should further state that the copy will remain in your possession at all times, and will never be left with another company for examination when you are not present. If the client is still concerned about confidentiality, then you should suggest one or more of the following additions, which usually remove any objections:

1. You will remove the company's name and any identifying information from the work used in the portfolio.
2. You will use only a representative piece of the work you create (preferably the refinement report), not the entire work.
3. The client has the right to approve the piece you choose.

In David Parkinson's article (mentioned earlier), he concluded by stating, "Remember, however, that the documentation product should also be seen as part of your project 'deliverables.' Left with the client, it can serve as your best 'business card.' Ensure that it is presented well and meets the clients' expectations. Along with the change control and custom operational procedures you have created for your client, you leave behind a professional account of your work."

You have now planned your work and worked your plan. You will gain impressive results by properly communicating your successful completion of the project during its refining stage. Like the craftsman who refines his precious metals by intensely heating the raw materials, you too can refine your project documentation by applying the practices covered in this chapter.

SUMMARY

Step 6 is refining your documentation. After the dust settles and everything has been done, you reflect back on your documentation from Step 5 and note what contributed to your success and/or failures in producing the desired objectives and deliverables. Basically, you go back and document what actually worked and what pitfalls are to be avoided next time. Your refined document should be a reflection of your best efforts in summarizing your work and the results of the project you worked on. This leads to the development of a knowledge management system whereby your knowledge, observations, training, and experiences are cataloged and used for future projects. Use common sense and best practices to avoid having your documentation used against you in the future.

The following exercises are located on N-Corp Services' web-site (http://www.n-corpservices.com) under the downloads section:

1. Project—Personal Computer memory hardware upgrade for 50 desktops. Using the Refinement template for external communication, provide a project summary report for this engagement.

Background: You work for a service company as a Computer Technician. You handle break/fix types of service calls, and you've been assigned a small project for an important account, the Smith Company. The project is to perform a personal computer memory hardware upgrade for 50 desktop systems, taking them from 128 Megabytes to 512 Megabytes of RAM. The manufacturer of the systems is Company X and they all the desktops are the same model and have the same operating system. You completed the project within scope and time constraints, and have already documented your project using the short-term documentation worksheet. You are now ready to present an external communication document to your customer to process the payment of your company's bill.

2. Project—Installation of a WIC module for a Cisco 3662 router. Using the external and internal communication template, write a technical summary for this project (as well as an internal edition) for your manager.

Background: You work as a Network Administrator for hospital that requires 24 X 7 up-time from its computer equipment. You've been assigned a project to add another clinic connection, Clinic A, to the existing main router. The T-1 circuits are now in place. The main router is a Cisco 3662 that already has the necessary memory, firmware version, and slots to accommodate the addition. The only thing you need to do is install a T-1 WIC (WAN interface card) module and at a later time you'll program the router and connect it to the previously-configured Clinic A router. You will need to power-down the router, install the WIC module, and turn it back on to perform the module insertion. Clinic A is due to be connected and ready to operate 20 days from the day you've been given this assignment. You are now on day 20 and have already completed the project. Clinic A has a PC/Network technician that will be supporting the site and who will require access to the router for troubleshooting purposes. Your goal is to write a refinement document for internal communication to be read by your manager. You will also need to write an external communication document for the Network technician who will provide maintenance support for this equipment. Please note that based on your previous experiences in working with the technician you feel he needs more training on WAN concepts and security to adequately support the site.

3. Project—Network infrastructure upgrade for Customer X. Provide an externally-targeted executive summary to the customer regarding this project, and in a convincing tone strive to be clear to the reader that your organization is best-suited to be their support vendor/partner. Provide another version (an internal communication) to those who may be called upon to provide the maintenance and support for this project, and address the concerns with management's handling of this project.

Background: You are a Network Engineer working for a services company that specializes in network design, support, and implementations. You've been assigned to a long-term project for Customer X that encompasses the design and planning of a network infrastructure upgrade from their existing 100-Megabit backbone to Gigabit Ethernet. The design will involve the purchase of new equipment and re-termination of their existing fiber optic backbone to handle Gigabit specifications. The project has taken roughly one month to complete (the initial timeframe was two weeks). You experienced various unexpected problems along the way. You had a crew of two network engineers and four network technicians assigned to this project. During the course of the project, your management consistently pulled out key engineers to work on other assignments. The network technicians lacked expertise on the product line that was sold to the customer. The customer was frequently not ready for your scheduled changes on their network and often had to reschedule cut-over dates and times. Despite the setbacks, the customer was pleased with the overall results of the project, but has reservations about seeking a service contract with your company.

4. Project—Remote site outage. Use the *Refinement* template for internal communication to report your findings to your support team.

Background: You are a Network Technician responsible for supporting remote sites within a 200-mile radius of the city in which your company is located. You have responsibility for three sites: the Finance and Accounting site, the Education and Training site, and the Marketing Department's remote office. Site 1 (Finance and Accounting) is located in City X, 45 miles from your corporate office. Site 2 (Education and Training) is located within your corporate office's main city—except it's on the far west side of town, about 25 miles from your corporate office. Site 3 (Marketing) is located near the corporate office, across the street in a business plaza. Each site has a LAN that connects to the corporate WAN via frame relay circuits ranging from 512K to a full T-1. You are only responsible for the LAN aspects of each site—the WAN support is done at your MIS Data Center by the Network Administrator. The most important site that you maintain is Finance and Accounting, since the CFO's office is located there and its business functionality is key to the success of your company. Your work hours are 8:00 A.M. to 5:00 P.M. You've resolved the issue of Site 3's outage, and you're now ready to report your findings to your support team for later review.

5. Project—Provide a report for external communication for a project you are currently working on. Provide an internal communication report to keep for future reference on this same project. The only reader of this report will be you.

CHAPTER 7

STEP 7—Maintaining & Updating Your Documentation

It's a common experience among IT professionals: Though you think you have a finished product and the assignment is complete, problems related to the project are encountered hours, days, weeks, or even months after your documentation is completed. This can be very frustrating and discouraging, especially since you've taken great care in documenting everything correctly, and you've painstakingly produced a great report during the refinement phase. It may seem that all that work was for nothing. However, keep in mind that without the proper documentation to fall back on you could find yourself redoing your work from scratch and from no reference point, which is a worse situation to find yourself in.

Another reason for maintaining your documentation is that you may need to bring in additional (or higher-level) technical support resources to review your work and to possibly make recommendations. One of the first things one does when engaging a customer on a project is ask for any current documentation. Unfortunately, most of the time the answer to this question is "We don't have any" or "It isn't complete or updated." Imagine how surprised these reviewers will be to see how thorough your documentation is. You may never know the depth of the impression it will have on those who review it. Your quality efforts may result in a promotion, job merit recognition award, pay increase, more projects, or a better job opportunity.

Here's a personal experience that happened to me as a customer reviewed the documentation binder I produced for a simple VPN (Virtual Private Network) router and firewall project. I was serving as the project manager and primary implementation resource on a small project for a healthcare insurance company. As you might guess, I was utilizing the *NuneX Method* of technical documentation for this project. Many of those in my service team (including the account manager) felt it was overkill on my part to thoroughly document this small project, but once the customer realized the level of detail and accuracy they could expect from then on, their confidence in our documentation processes resulted in more business opportunities and future project engagements. As it turned out, they were testing us with a small project in order to engage us on larger ones. As we obtained more service business opportunities, my documentation binder was used as the basis for continuing the work we started into a network infrastructure rollout. This aspect was added to the Maintenance phase of the documentation binder since it built on work previously done. Some may argue it was overkill to do this level of documentation for a $2,500 project, but in a three-month period it resulted in more than $100,000 in business revenue due to the high level of confidence they had in our documentation and quality processes. I later used the documentation binder for this small project as a selling tool to obtain service business opportunities with other companies. No project is too large or too small for the *NuneX Method*. It works for any project or engagement, regardless of project timelines.

Having your documentation nicely organized into the *NuneX* 7-steps can do the following for you:

1. It will indicate that you've followed a systematic approach to your documentation.

2. It demonstrates that you have given the project serious thought and have devoted your attention to detail.

3. It will allow the customer to follow in your footsteps—and quite possibly help you determine where the causes and affects around an issue may lie.

4. It conclusively shows that you didn't take an unorganized and hurried approach to the assignment, but instead utilized the 80/20 rule to achieve efficiency and success.

5. It will prove that you are current and up-to-date on all of your projects—and that you have the documentation to prove it.

6. It will generate customer confidence in your abilities to start and finish a project and to deliver impressive results.

7. It may also result in better job opportunities as other potential clients review your technical work or portfolio.

Updating your documentation can help you provide the feedback that is crucial for a procedure or process to be improved. Maybe while you're updating your documentation you realized or discovered a whole new approach to a procedure. A system requires feedback in order to improve the internal processing that goes between the input and output. The 7th step in the *NuneX Method* is your documentation's feedback process, which serves the purposes of maintenance and updating. Even if you don't produce the desired results in your endeavor, at least you have documented a way with which to **not** do it. Thomas Edison once wrote, "Genius is 1 percent inspiration and 99 percent perspiration."

The drawings that follow illustrate how the *NuneX* 7th step—Maintenance—serves as a feedback mechanism for a completed project.

System

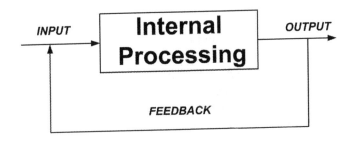

Nunex Method

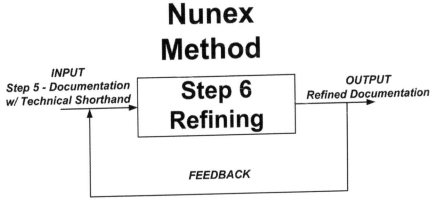

Some projects may be completed and closed when you finish refining your documentation. You may never have to revisit the issue again. However, for the sake of quality, you may want to do a follow-up on several projects to confirm your successes. We often say, "No news is good news." However, "no news" could also mean "We've been too busy to let you know it didn't work." It is wise to follow-up and update files (even via a random sampling) just to get a good feel for the quality level of work done on completed projects.

How is the maintenance phase performed using the *NuneX Method*? Do we go back and update every section of our documentation binder or system? The answer is no; you create a new section called "Maintenance" where you provide any updates to any step of the *NuneX Method*. In other words, updated information is kept together in step 7—since in step 6 you already produced the final document. There is no need to produce that information again; you just keep updating it in a separate section. You utilize the same worksheets you used in the previous six steps—the only difference is that you'll compile these in the Maintenance section of your documentation binder. If your documentation is updated often enough you ought to consider producing an updated refinement follow-up document and sending it to the stakeholders. You can use the Refinement template for external and internal communication—but be mindful about communicating that it contains updated information.

Another application for the *NuneX* Maintenance phase of technical documentation is updating and documenting changes for network diagrams, addresses, applications, and any other pertinent information. This will allow you to have the most accurate and up-to-date information available regarding your customer and systems.

One of the most frustrating assignments can be to take over a project someone else started—you may feel the project is set for failure due to the lack of attention your predecessor devoted to it. By using the *NuneX Method* you can quickly come up to speed on a project and be able to successfully meet its objectives and deliverables. Your efforts to document and follow the first six steps of the *NuneX Method* will allow you or someone else to easily maintain it thereafter—a rare win-win situation in an IT shop.

SUMMARY

Step 7 is maintaining and updating your documentation at a point well after the project is completed or the problem is resolved. You'll want to keep updating the results of your work, as sometimes problems have a cause and effect basis that may take time to reveal. Having a recorded accounting of events after you perform the project can help you refer back to provide clean-up work or to analyze data. Maintenance can also be a quality check on your work as you follow-up with those whom your project affected.

Exercises (answers are located on N-Corp Services' web-site, http://www.n-corpservices.com, under the downloads section):

1. **Project—Personal Computer memory hardware upgrade for 50 desktops. Use the Security worksheet you used for Step 3 to update the altered power-on passwords in the Maintenance section of your documentation binder.**
Background: You work for a service company as a Computer Technician. You handle break/fix types of service calls, and you've been assigned a small project for an important account, the Smith Company. The project is to perform a personal computer memory hardware upgrade for 50 desktop systems, taking them from 128 Megabytes to 512 Megabytes of RAM. The manufacturer of the systems is Company X and all the desktops are the same model and have the same operating system. You completed the project within scope and time constraints and have documented your project using the *Short-term Documentation* worksheet. Some days after the project was completed your company receives a call that one unit's power-on password is not allowing access to the system. You are then dispatched to find out what is wrong. As you troubleshoot the issue, it turns out there was a computer swap between two managers of the company. You reset the power-on password for both units to reflect correct operation. Update the appropriate forms on the Maintenance section of your documentation binder.

2. **Project—Network Infrastructure Upgrade for Customer X. Using the refining documentation provided on the project, propose the best solution to solve a redundancy problem using the *Planning* and *Contingency Planning* worksheets.**
Background: You are a Network Engineer working for a services company that specializes in network design, support, and implementations. You've been assigned to a long-term project for Customer X that encompasses the design and planning of a network infrastructure upgrade from their existing 100-Megabit backbone to Gigabit Ethernet. The design will involve the purchase of new equipment and

re-termination of the existing fiber optic backbone to handle Gigabit specifications. The project has taken roughly one month to complete, but the initial deadline was two weeks. You experienced various unexpected problems along the way—including personnel issues. You had a crew of two network engineers and four network technicians assisting you with this project, but during the course of the project, management consistently pulled out key engineers to work on other assignments. The network technicians lacked expertise on the product that was sold to the customer. Added to this, the customer was frequently not ready for your scheduled changes on their network, and you found yourself having to reschedule cut-over dates and times. Despite these setbacks, the customer was pleased with the overall results of the project, but has reservations about establishing a service contract with your company. Some days later, as you proceed with service contract negotiations with your customer, one of the core switches goes down, and the designed redundancy failure mechanisms for the backup core switch fails. Use your documentation binder to locate the point of failure and provide alternative solutions.

3. Project—Update the Maintenance section of your documentation binder for one of your recently completed projects. If the information updated is significant, rework your refinement document.

CHAPTER 8
Putting IT All Together

You've just completed reading about the seven steps of the *NuneX Method*. You've learned what's involved in each step, from Pre-Documentation to Maintenance, and how each step builds on the one before it. The first four steps of the *NuneX Method* (Pre-Documentation, Planning, Security, and Notation) are part of the 20 percent preparation efforts of a project that produce 80 percent of the viable results. The remaining three steps (Documentation, Refinement, and Maintenance) produce the remaining 20 percent of results that lead to the successful completion of a project marked by quality work, knowledge management, and impressive outcomes.

Having reached this last chapter, I hope you've become energized and ready to apply the information presented in this book to your everyday technical work. Let me set some real expectations from this point on: IT professionals (especially those who do service work) are extremely busy individuals who usually work in short-staffed and fast-paced IT shops. It *will* take some time to get accustomed to using the *NuneX* methodology. You may not be able to apply it overnight. My recommendation to you is to learn the concepts of the seven steps and memorize the technical shorthand table and security codes—or even create your own versions. You may be tempted to skip some steps or hurry through others. However, if this method is to contribute to your success as a technical resource, you must follow it step-by-step, in its entirety. Believe me, investing in the little things will pay off with big ones. The old saying "practice makes perfect" is replaced by this saying: "*perfect practice makes perfect.*" To perfect the *NuneX Method*, you must work on the mastery of its subcomponents until they become easy and logical for you to follow.

133

Imagine a beginner in the sport of archery as he learns to hit a target. Hitting the bull's eye on the target with the arrow is considered to be the successful completion of the archer's objective. There are many things to consider while shooting the arrow, such as the state of the equipment being used, the archer's stance, the target's range, aiming technique, the archer's breathing, pulling back the bowstring, safety procedures, and finally the follow-through of shooting the arrow. Each aspect of shooting the arrow builds on the mastery of each prior subcomponent to ensure correct handling and proper execution. After much repetition, a beginner will perform all these steps without conscious thought—and if training, practice, and progressive mastery of the right techniques are applied he will consistently hit the target. It doesn't happen overnight, but the successful completion of the objective requires attention to every detail and continued practice of execution and delivery. As each arrow is released and then analyzed for possible improvement, mastery will eventually be achieved. If an arrow is shot without regard to the necessary techniques for achieving a successful result, the archer's efforts will be inconsistent and he will not regularly hit his target. The archer may even become a safety hazard to himself and others. Any beginning archer is capable of learning to shoot a bow and arrow on his own after a lot of trial and error, but he will certainly learn these techniques faster and better from an experienced archer that teaches the concepts and techniques needed to properly shoot. As the objective becomes more challenging and complex—such as hitting a moving target—a beginning archer's mastery of the fundamentals becomes even more essential.

The application and execution of the *NuneX Method* is similar to the work of the archer. Without following the necessary steps that build on each other, successful completion of a project and the meeting of set objectives and deliverables will usually not be accomplished. Our goal is to consistently hit our intended target by paying attention to the details that will assure impressive outcomes. Like the beginning archer who can develop mastery of the sport by being taught by an experienced professional, your application of the *NuneX Method* has been taught to you by an experienced documenter who has shown you proven techniques needed to achieve success. As you embark on a new project, imagine yourself as the archer aiming to get a perfect shot at the bull's eye. Via the successful completion of a project using the *NuneX Method*, you will achieve the same effect as an experienced and first-class archer hitting a target. Remember, an outstanding job is seen and felt by everyone affected by your work.

The most challenging aspect of the *NuneX Method* is becoming proficient in technical shorthand and developing your own security-coded system. All the concepts and material presented within this book are based on best practices, logical process flow, years of experience in the Information Technology field, and common sense.

This method was developed by an IT professional concentrating on technical specialties *for IT professionals*. It has been tested in various areas within the technical realm of IT, such as project management, network administration and engineering, hardware support, system implementations, troubleshooting, and even PC service and troubleshooting. In all these cases (and many more) the *NuneX Method* has successfully contributed to the success of those who tested this system and provided feedback on its refinement.

The best approach to learning the *NuneX Method* is to first learn the manual system, then (after a few projects have been completed) starting to document projects electronically. It is important to consistently keep the documentation in one form so you can easily reference all the information contained within it. In other words, don't mix electronic information with paper information. Print out the electronic documentation and put it in your documentation binder if you are using the manual system. "Doing it manually" means to utilize the pen-and-paper approach of documenting your work (a more descriptive and detailed list follows). "Doing it electronically" means to utilize the various hardware and software tools available, which range from a simple e-mail client software application to a single-user database program to a complex internet/intranet multi-user application.

To apply the *NuneX Method* manually, you can use common office supplies to organize various sections. I suggest (as a minimum) the following materials:

- One (1) three-ring notebook binder. Use a 1.5" binder for your immediate and small-scale projects and a 3" (or larger) binder for long-term projects or for transferring inactive projects or documentation from the 1.5" binder. It's a good idea to keep as much documentation in the 1.5" binder as possible—you may opt for a 2" binder. A 1.5" binder is easier to carry with you to meetings, remote sites, and home for on-call responsibilities. After some time of inactivity in a project, you can archive your information in a larger binder.
- Eight (8) section dividers. Seven are used to divide the seven steps of the *NuneX Method*, and the eighth can be used for miscellaneous information. The first will be used for "Pre-Documentation" information. The second section will be "Planning," the third will be "Security," the fourth "Notation," and the fifth section is the actual project documentation. The sixth section will be your final "Refinement" production, and finally, the seventh section will be the Maintenance and updating section. I use the last section for financial records, receipts, or items within my documentation that require quick access. You may also choose to place any communication logs, maps, and even the technical shorthand reference sheets in the eighth section for quick access. A ready-to-go solution for the divider requirement would be to use the *Avery Ready Index 8-Tabs* (PN: RI2138—11133) commonly found

in most office supplies stores. Another Avery product that works well is PN: 11817 *Ready Index 8-Tabs with Translucent color*, if you want a more fancy presentation format. The layouts for these dividers can be used with Microsoft's *Word* or Corel's *WordPerfect* word processing software. You can download the free Avery Wizard software application or templates from the Avery web-site to print text on these dividers. Another readily available solution can be found by purchasing a Wilson Jones view tab transparent divider 8-tab multi-color index package. You can also download templates from the Wilson Jones web-site. Some of my peers prefer to use the Wilson Jones solution because its organization lends to quick and easy customizable tabs, without the need to print labels, tab inserts, or special paper—it's a one-step format for creating title pages and tables of content. By the time this book gets published and into your hands, both manufacturers will have updated these part numbers with better organizer/divider solutions, so my recommendation is to browse through your favorite office store to find a solution that suits your taste and needs.

- Pens and pencils for writing and drawing diagrams. It is preferable to use a mechanical pencil (0.5mm) and a fine-point black pen.
- A highlighter pen for note-taking.
- A hole-punch for papers that need to be put in your binder—ideally a three-hole model. You may want to invest in a heavy-duty hole-punch if you plan to utilize the manual system as your standard documentation procedure.
- Pre-printed Worksheets for each step in the *NuneX Method*. The pre-printed worksheets to include in each section are:

Table of Contents Section (Already included with the binder tab solution)
    NuneX Method Flow Chart

Pre-Documentation Section
    Internal/External References Worksheets
    Pre-existing LAN/WAN Diagrams
    Project Plan (if provided by a project manager)
    Scope of Work (if written by someone other than you)

Planning Section
    Planning Worksheets
    Meeting Planning Worksheets
    Technical Project Status Meeting Agenda Worksheets
    Contingency Planning Worksheets
    Rules of Engagement Worksheets
    *Travel Maps of site locations
    *Travel itineraries, flight schedules, and hotel information
    Scope of Work & Project Plan if written or updated by you

Security Section
> Security Worksheets

Notation Section
> Printed notes and information from external resources

Documentation Section
> *Communication Log Worksheets
> Short-term Project Documentation Worksheets
> Documentation Worksheets
> Note-taking Worksheets
> Meeting Attendees Worksheets
> Meeting Minutes Worksheets
> *Submitted and Processed Company Expense Forms
> Note-taking Worksheets for knowledge gained while working on projects.

Refinement Section
> Status Reports
> Project Plan Updates
> Refinement Documentation Template
>> —External Audience Communication
> Refinement Documentation Template
>> —Internal Audience Communication

Maintenance Section
> All previously listed worksheets are applicable.
> If the project is significantly updated, then rewrite Refinement documents for both internal and external communication.

Miscellaneous Section
> Optional: all Worksheets preceded by an asterisk (*) character
> Financial documents, receipt copies, etc.
> Quick reference information such as Technical Shorthand Table

Appendix A includes these worksheets for your use. You may also download the same or updated electronic versions from N-Corp Services' web-site, http://www.n-corpservices.com , under the *Downloads* section.

How to use the binder

If you will be sharing your project documentation with others in a team, use a 3" or 5" binder as the master repository, and a 1.5" or 2" binder for your personal records. Always be sure to make double copies of any information in your binder—both in the master repository and in your personal binder.

The following is a list of steps to organize your project binder for use with the *NuneX Method*:

Step 1: Start by downloading the free Avery wizard at http://www.avery.com to work with your favorite word processor. Install and run the wizard for the "ready index" product. Be sure to use the biggest font size possible for easy viewing. Alternately, you can use the Wilson Jones ready-to-use templates in MS Word or Word Perfect format, found at http://www.wilsonjones.com.

Step 2: After installing the Avery wizard software or the Wilson Jones template document, run the Avery wizard application or type in the pre-set fields in the Wilson Jones template document to create the following list for your table of contents:

1. Pre-Documentation
2. Planning
3. Security
4. Notation
5. Documentation
6. Refinement
7. Maintenance
8. Miscellaneous & Quick Reference

Step 3: Make copies of the worksheets (Appendix A) or print electronic copies from the download section of N-Corp Service's web-site, punch them for your 3-ring binder, and place them in the appropriate sections.

Step 4: Starting with section 1, Pre-Documentation, complete each worksheet as it corresponds with the actual phase in your project. Fill out each section as you progressively utilize the 7 steps of the *NuneX Method* for technical documentation and knowledge management. You may jump ahead several steps in the *NuneX Method* as you obtain information relative to each phase of your technical project. It is important that all your worksheets contain the appropriate naming convention for each project and/or meeting for consistency.

Step 5: Upon completion of your project, develop your refinement report for both internal and external communication. Keep a copy of it in section 6 of your binder. It is recommend that you formally present your refinement reports to the project stakeholders and intended audience with separate binders paying attention to aesthetics, flow, clarity, and include any electronic copies of pertinent files within a CD-ROM or DVD. This separate binder can serve as "turnover documentation" upon the project's completion.

Step 6: Add any new updates or additions to your project to section 7, Maintenance.

Step 7: If there is a period of inactivity for this project, transpose the contents of your project binder to an archival binder with enough space to accommodate it.

It is important to *never* remove original pages from the project binder that you're using as your centralized documentation repository without making copies; these pages may be lost or misfiled. Keep a floppy disk, CD-RW, or DVD-RW in the project binder and copy the electronic versions of documents in it so they can be reprinted or updated when necessary.

Going from manual to electronic documentation

When you have completed the project in a manual format, you may then store the associated electronic documents (using the NuneX project naming convention) in sub-folders correlating to each step of the *NuneX Method*. The folder hierarchy can then be saved to a CD-ROM or DVD by using a Burner software application. Be sure to use CD label printer paper with a color inkjet to print a CD label that contains the project name, project start and ending dates, and name of the documenter. You can store each project in a separate binder containing CD holder slots (or a CD carrying case) to keep track of each project you have completed electronically.

Here's how you may archive your files electronically:

Root Folder—Name of Project

    Sub-folder 1—Pre-Documentation

    Sub-folder 2—Planning

        Sub-folder 2 A—Contingency Planning

    Sub-folder 3—Security

    Sub-folder 4—Notation

    Sub-folder 5—Documentation

    Sub-folder 6—Refinement

        Sub-folder 6 A—External Communication

        Sub-folder 6 B—Internal Communication

    Sub-folder 7—Maintenance (please note that if you create a read-only CD-ROM you won't be able to add updated files; it is better to utilize a CD-RW for allowing updates to be made on the disc)

Some documenters prefer to do their electronic "turnover" documentation in a web-page format using popular web-design products (such as Microsoft's *FrontPage*) to create a web-page layout of their documentation. Turnover documentation is documentation that is turned over to a customer, manager, team, or other designated entity that needs to have a copy of your documentation. For most applications, a copy of your Refinement document is usually adequate. For your own copy, we suggest you include *all* your documentation. Visit our website at http://www.n-corpservices.com for an example of turnover documentation that was created in a web-page format.

Hey, we're IT guys! Isn't there an easier way to do this?

It is only fitting that since we are IT professionals we'd have a computerized way of utilizing the *NuneX 7 step Method* of documentation. Yes—there are various ways to document electronically. The easiest and most inexpensive way is to use your current e-mail application to store files. The main reason for utilizing e-mail is that most project work involves the communication, dispensing, and gathering of documents—usually sent electronically via e-mail. For those hardcopy documents that you wish to store electronically, you can utilize a scanner with optical character recognition (OCR) capability or use the poor man's scanner technique (i.e., fax it to yourself), if you have a fax server or application that allows you to receive faxes on your computer.

To utilize the electronic system effectively, you'll need to organize your folders and files in the best way for easily retrieving and adding new information. The *NuneX* methodology lends to this organization process by following the naming convention and 7 steps.

Here's an example of how you may organize your documentation electronically using an e-mail application front end:

INBOX—Folder (Project Name using NuneX project naming convention)
Sub-Folders to Inbox "Project Name" folder:
　　Pre-Documentation
　　Planning
　　Security
　　Notation
　　Documentation
　　Refinement
　　Maintenance
　　Miscellaneous

If you can save these files on your company's main mail server you may decide to share these with your team in a public folder, where all projects can be electronically stored for easy access. Be sure to invest some time in planning security and archival for these folders to allow yourself or your team to fully edit your assigned projects, and others to have view-only access.

There are various options for electronically storing your documentation files. We are currently in development on an easy-to-use Microsoft Access program (which can be downloaded from N-Corp Services web-site), a pocket access Windows CE application version that can be later used in conjunction with popular PDAs (Personal Digital Assistants), as well as intentions for producing a new web-site (to be later announced at http://www.n-corpservices.com) where you can securely store your technical documentation following the *NuneX Method*, for a nominal monthly or yearly fee. Other electronic documentation solutions will be provided as the methodology progresses and feedback is provided to improve each product.

What's next?

As you become proficient in using the *NuneX Method*, there is one last piece of advice and information that I wish to impart to you. If you have the discipline and energy to do so, and especially if you are in a technical management position, it is important that you **document everything that you feel needs to be referenced!** What does this mean? I mean you must document everything from phone calls, voice-mails, expenditures, faxes, correspondence, to projects that you don't own but participate in—anything and everything that may be relative to your work in one way or another. Why do this if we are already using the NuneX Method to document our work, which already includes some of these tasks? *We live in a political society that focuses it energies on placing blame and seeking accountability rather than fixing problems.* Why are there so many lawsuits? Why do people get fired for missed opportunities and for mistakes? Beside the obvious reasons (such as gross negligence and lack of industriousness) the fact that most people don't document their activities and experiences eventually becomes their downfall. A good way to view total and accurate documentation is as though it were your personal guardian and protector. You need to take care of the people and things that can and will take care of you. Documentation is the resource you can count on *as long as you put in the effort and maintain it.* Documentation is, without a doubt, a time-consuming task. However, as the 80/20 rule implies, the 20 percent time invested in this pursuit will return 80 percent rewards, one way or another. It will better prepare you to handle anything and everything that may come your way, be it personal or professional.

Conclusion

It is my hope and expectation that the information presented in this book will help you become a better technical documenter and IT Professional. Your success in utilizing the *NuneX Method* to become the best at what you do will be my success as an educator and developer of the methodology. Like the master archer who consistently hits his intended targets, you can achieve the same mastery of the *NuneX Method* with practice, persistence, and application of the technical documentation techniques presented in this book. The *NuneX Method*—in a nutshell—is planning your work, working your plan, and systematically documenting your progress along the way.

If you wish to contribute to the information pool of technical documentation and knowledge management best practices contained within the *NuneX Method*, please feel free to provide us with feedback. We'd like to hear about your personal successes in utilizing the *NuneX Method*. Feel free to write to us by sending an e-mail to: info@n-corpservices.com. Please be sure to provide us with your name, address, and phone number, as well as your feedback on the *NuneX* methodology, and its impact on your work environment. You may also visit our web-site at http://www.n-corpservices.com and give us feedback at any time.

Exercises

1. Put together a manual binder with the associated worksheets and organize it in its various sections.
2. Create a documentation binder regarding the project you have been documenting using the *NuneX Method*.
3. Transpose your manual documentation to an electronic documentation CD-ROM using the *NuneX Method*.
4. Create a web-page report using the Refinement documents with links to each step of the *NuneX Method*.
5. Create an e-mail sub-folder hierarchy that follows the *NuneX Method* for one of your current projects.

APPENDICES

NUNEX METHOD WORKSHEETS

The nun=x Method
Technical Documentation Techniques

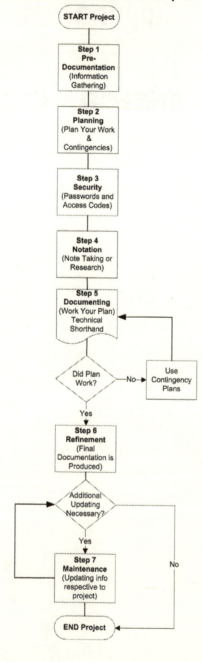

Internal / External Reference Worksheet

Modification Date: _____

Position	Name	e-mail address
Phone		

Legend

IR	= Internal Resource	Sv	= Supervisor	T	= Technician
ER	= External Resource	S	= Systems	Op	= Operations
Ex	= Executive	H	= Help Desk	Pg	= Programmer
FTE	= Full Time Employee	A	= Analyst	G	= Consultant
PTE	= Part Time Employee	L	= Leader	D	= Director
nwc	= No longer with Co.	na	= Not Available	Pc	= Personal Computer
				N	= Network
				M	= Manager
				P	= Project
				C	= Chief
				O	= Officer

Ref #	Name	Phone	EXT	Cell Phone Pager	e-mail address Work & Home	T Zone Hrs Days	On-Call	Home Phone	Home Address	EMP Status	Super-visor

CODE	Meaning
INST	Installation
UPG	Upgrade
HW	Hardware
SW	Software
NET	Network
WAN	Wide Area Network
LAN	Local Area Network
DES	Design Project
SC	Support Call
PRB	Identify Problem Only
REP	Repair
SET	Setup
CON	Configuration
TRB	Troubleshoot
SRV	Service
SI	Support Issue
PRJ	Project (long-term)
FW	Firmware
COM	Communications
FW	Follow-up
IMP	Implementation
MAINT	Maintenance

PLANNING WORKSHEET

DATE: ___ / ___ /200__ DAY _____ TIME: ____:____ AM PM PG ____ of _____ PGS
Project Name:
Planner(s):
Projet Description:
Projet Scope & Set Objectives
Projet Deliverables

Project Plan Sequence											

Plan Results
Plan Completion Date: ___ / ___ /200___ DAY _____ TIME ____:____ AM PM

MEETING PLANNING WORKSHEET

MTG Date: _____/_____/_____ Day: _____

Meeting Location: _____

MTG Facilitator(s):_____
_____

Will Attendees need directions to meeting location? _ Yes _ No

Needed Resources:
_ Overhead Projector   _ Whiteboard w/markers
_ Overhead Computer Display   _ Flipper Board w/markers
_ Speakerphone Capability   _ Laptop Computer for Presenting
_ Agenda Sheet # _____   _ Writing Materials (Pen & Paper)
_ Project Status Sheet # _____   _ Attendance Sheet # _____
_ Network Access—Security _Y _ N   _ Internet Access (Browser _____)
_ Tape Recorder w/tapes   _ Other: _____

Setup Time: _____ _ AM _ PM   Responsible _____
Start Time: _____ _ AM _ PM   End Time: _____ _ AM _ PM

Reference to Previous MTG Minutes Sheet #:_____
Change to MTG Format as designated by last MTG's Minutes
_____

Meeting's Purpose
_____

MTG scheduled and all attendees notified? _ Yes _ No
Responsible _____

TECHNICAL PROJECT STATUS MEETING AGENDA WORKSHEET

MTG Date:_____/_____/_____ Day:_____

Meeting Location: _____

MTG Facilitator(s):_____
_____

MTG Recorder/Secretary: _____
Start Time: _____ _ AM _ PM   **End Time:** _____ _ AM _ PM
Reference to MTG Planning Worksheet #:_____
MTG Rules _ New Set _ Addendum _ Revision

AGENDA
I   **Administrative**
   _ A. Review of Previous MTG Minutes
   _ B. Policy Changes
   _ C. Funding Issues
   _ D. Project Status/Updates
   _ E. New Assignments
   _ Other _____

II **Team Member Status Updates**
   _ A. Tasks previously worked on
   _ B. Tasks currently working on
   _ C. Next project or assignment
   _ D. Issues that are creating problems
   _ E. What the team can do to provide assistance
   _ F. Timelines
   _ Other _____

III **Improvements/Suggestions to MTG formats**

IV **Meeting Minutes Updates Info.**
   Dispensed via _ e-mail _ interoffice mail _ postal mail _ fax _ network
   _ Other _____ _ all of the above
   _ Date: _____
   _ Responsible: _____

V   **Conclusions/Review/Adjourn Meeting**

CONTINGENCY PLANNING WORKSHEET

DATE: ___/___/200__ DAY ___   TIME: ___:___ AM PM   PG ___ of ____ PGS

Project Name:

EXTERNAL SUPPORT RESOURCES

Resource Name:
Title:
Company Name:
Phone:
Cell:
FAX:
Pager
e-mail
Customer #
Other:

Resource Name:
Title:
Company Name:
Phone:
Cell:
FAX:
Pager
e-mail
Customer #
Other:

INTERNAL SUPPORT RESOURCES

Resource Name:
Title:
Department Name:
Phone:
Cell:
FAX:
Pager
e-mail
Customer #
Other:

Resource Name:
Title:
Department Name:
Phone:
Cell:
FAX:
Pager
e-mail
Customer #
Other:

Backup Day	Sun	Mon	Tue	Wed	Thu	Fri	Sat	Other
Backup SET								
Backup Date								

CONTINGENCY PLAN A

CONTINGENCY PLAN B

Rules of Engagement Worksheet
Best Practices to Consider for Planning Your Projects

Rule #	Check	Rule Summary
1		Secure your area, systems, & networks
2		Protect your organization against harmful agents & viruses
3		All systems should be fully redundant
4		Plan for growth
5		Plan for training
6		24/7 Support for HW, SW, & Networks
7		100% capable backups – test disaster recovery plans
8		Document your operational processes
9		Have adequate power for current & future growth
10		Monitor all of your systems – be proactive rather than reactive
11		Have a separate test & production system
12		Standardize your tools & materials needed to complete the job
13		Use well supported technologies
14		Effectively manage licenses & audit hardware
15		Hire competent staff – set high standards
16		Don't talk to salesmen unless you have money to spend or are preparing a budget
17		Consider all possible options & be open to new ideas
18		Talk to the experts regarding product selections
19		Keep the environment of the equipment within ideal operating parameters
20		Train the Help Desk for support calls regarding your projects
21		Don't hold back information from those who need to know
22		Always thank those who help you
23		Plan for cabling
24		Eliminate human dependencies
25		Don't place blame, get problems fixed
26		Don't start until everything is ready
27		Be prepared to ship equipment back
28		Factor additional costs for a project
29		Multiply your estimated time for a project by a factor
30		Plan your site surveys & avoid assumptions
31		Require a Scope or Statement of Work proposal along with a project plan from any 3rd party you engage for contract
32		Quality, Speed, & Price
*****		Get everything in writing – all communications, correspondence, etc.

Sample Code Section Table:

Alphanumeric Character	Sample Code
0	ϒ
1	'
2	≤
3	/
4	∞
5	*f*
6	Ω
7	Ψ
8	Ξ
9	α
10	↔
A	←
B	
C	→
D	↓
E	∘
F	±
G	″
H	≥
I	×
J	Π
K	⇔
L	•
M	÷
N	≠
O	≡
P	≈
Q	…
R	⌶
S	—
T	⌡
U	⊗
V	⊕
W	∅
X	∉
Y	⊂
Z	⊇

Security Worksheet

Page_____   Project _____

Documenter(s): _____

_____

Source: _____

_____

System	Account Name	Password

DOCUMENTATION WORKSHEET

DATE: ____ / ____ /200__ DAY _____   TIME: ____ : ____ AM PM   PG ____ of ____ PGS

Project Name:

Documenter(s):

PROCEDURES   **Defined Variables:**

Procedures		
01	31	51
02	32	52
03	33	53
04	34	54
05	35	55
06	36	56
07	37	57
08	38	58
09	39	59
10	40	60
11	41	61
12	42	62
13	43	63
14	44	64
15	45	65
16	46	66
17	47	67
18	48	68
19	49	69
20	50	70
21		
22		
23		
24		
25		
26		
27		
28		
29		
30		

Note Taking Worksheet

DATE: ____/____/200__ DAY _____ TIME: ____:____ AM PM PG ____ of ____ PGS

Project Name:

NOTES

Information that is received   **Information that is dispensed**

Diagrams / Charts / Drawings

TECHNICAL MEETING ATTENDEES WORKSHEET

MTG Date: _____/_____/_____ Day: _____

Meeting Location: _____

MTG Facilitator(s): _____

_____

Start Time: _____ _ AM _ PM   End Time: _____ _ AM _ PM

Reference to MTG Planning Worksheet #: _____

Name	Company	Title / Position	Signature

MEETING MINUTES WORKSHEET

MTG Date:_____/_____/_____ Day:_____

Meeting Location:_____ MTG Recorder/Secretary: _____
Start Time: _____ _ AM _ PM   End Time: _____ _ AM _ PM

Reference to MTG Agenda Worksheet #: _____

_____   _____
MTG Recorder/Secretary's Signature   Date

MTG Rules _ New Set _ Addendum _ Revision

Agenda Item / Presenter	Discussion/ Conclusion	Recommendation/ Action	Follow-up/ Responsible Party
Administrative			
Status Updates			
MTG Improvements			
Conclusions			

Attendees List Sheet #: _____
Next Meeting Time :_____ _ AM _ PM
Location: _____

Short-Term Project Documentation Worksheet

DATE: ___/___/200__ DAY _____ TIME: _____:_____ AM PM **PG** ___ **of** ____ **PGS**

Project Name:	Start Date:
Documenter(s):	Completion Date:

Customer Name: _____	Contact Name: _____
Customer Address: _____	Contact Phone: _____
_____	Contact Cell / Pager: _____
City _____ ZIP _____	Contact e-mail: _____
Work Hours _____ Time Zone _____	After Hours Contact: _____

Ticket or Case #:	Internal	External

Problem Description or Service Request

_____
_____
_____

Steps 1 - 3

Steps 4 - 6

Documentation Process:

Step 1: Verify Information	Step 2: Equipment Information	Step 3: Answer H5W for Request
Step 4: Answer H5W for Resolution	Step 5: Observations	Step 6: Miscellaneous

H5W: Who, What, When, Where, Why & How

Completed By (Print Text):	Signature:
Work Acknowledged By:	Signature:

Communication Log Worksheet

Project: _____ Page ___ of ___

Name of Company or Individual Called R=Received call from C=made Call to V=Voicemail	Phone Number Dialed P=Phone C=Cell PGR=Pager ~ = Approximately	Date & Time A=AM P=PM	Result B=Busy LVM=Left V-Mail LMwP=Left Message w/Person NR=No Response

Refinement Documentation Template

External Audience Communication

Outline:

Report Date, Day, & Time

Name of Report Writer

Introductory Preview

Project Background & Scope (include the business needs for this project)

Project Objectives & Deliverables

Project's Financial Information (revenue, expenditures, & return on investment)

Project Timeline

Project Team Participants (with department names, e-mail, dates, & phase participation)

Project Issues, Transitions, & Key Observations

Project Results

Project Conclusion Statement/Summary

(Knowledge Management Aspects)

For Internal Audience Communication add the following:

1. What aspects of the project worked well and achieved desired results (i.e., did plan work?)

2. What aspects of the project didn't achieve desired results (i.e., did you need to use contingency plans?)

3. What did you personally learn by working on this project?

4. Your personal observations of people, time constraints, resources, equipment, software, tools, etc.

5. If you had to do this project all over again, what would you do differently?

6. An overall synopsis or summary of your version of the project—the whole truth about it.

BIBLIOGRAPHY

Covey, Stephen R. (1990). *The 7 Habits of Highly Effective People. New York:* Simon & Schuster.

Freedman, Rick. (2000), *The IT Consultant: a commonsense framework for managing the client. San Francisco*: Jossey-Bass Pfeiffer.

McClure, Stuart, Scambray, Joel, Kurtz, George.(1999), *Hacking Exposed: Network Security Secrets & Solutions. Berkeley:* Osborne/McGraw-Hill.

Arrredondo, Lani (1991), *How to Present Like a Pro.* Baskerville: McGraw-Hill.

Haramundanis, Katherine (1998). *The Art of Technical Documentation 2nd Edition. Boston:* Digital Press.

MaGuire, Francis X., (2001), *You're the Greatest!: How Validated Employees Can Impact Your Bottom Line. Germantown*: Saltillo Press.

Disney Institute—Seminar on Customer Service—April 2000

Cleland, David I, ed., Field Guide to Project Management. John Wiley & Sons, 1998.

Hallows, Jolyon. Information Systems Project Management: How to Deliver Function and Value in Information Technology Projects: American Management Association 1998.

Kim Heldman, PMP, PMP: Project Management Professional Study Guide 2nd Edition, Sybex 2004.

Lientz, Bennet P., Rea, Kathryn P., Breakthrough Technology Project Management, Academic Press, 1999.

Murch, Richard, Project Management: Best Practices for IT Professionals, Prentice Hall, 2000.

Taylor, James, A Survival Guide for Project Managers, American Management Association, 1998.

Project Management Institute, Inc., A Guide to the Project Management Body of Knowledge (PMBOK® Guide), 2000 Edition.

Web-Sites:

TechRepublic.Com

AskMe.Com

Techies.Com

WhatIs.Com

CIO.Com

PMI.Org

Tenstep.Com

Drj.Com

MPUG.Org

N-CorpServices.Com

ABOUT THE AUTHOR

Richard Nuñez is the founder and developer of the NuneX Seven-Step Method for Technical Documentation Techniques and Knowledge Management. He has worked in the computer service and support industry for more than 14 years, and has successfully progressed from an entry-level service technician to an executive manager. His pursuit of success in the Information Technology field was the driving force behind the development of the NuneX Method.

Mr. Nuñez has achieved professional certifications in various operating systems and products, such as Project Management, Novell, Microsoft, CompTIA, Cisco, Citrix, and Help Desk (among others). He is also a member of the Border Network Chapter of the Society for Technical Communication, Project Management Institute, Association of Information Technology Professionals, and the Society for Documentation Professionals. Mr. Nuñez has an Associate's Degree of Occupational Studies from Western Technical Institute (WTI) in El Paso, Texas, in Microcomputer Technology. He also has a Bachelor's Degree of Science in Business and Information Systems from the University of Phoenix (UOP).

Mr. Nuñez has worked in the private and public sectors, specializing in technical areas such as personal computer service and repair, network administration and engineering, and technical supervision and management. He has an educational background in electronics, computer technology, network administration, network engineering, and business management, and has taught technical courses and seminars related to technology and documentation techniques.

Mr. Nuñez currently serves as the Chief Executive Manager for N-Corp Services, LLC where he writes about, enhances, and promotes the use of the *NuneX Method*. He lives with his wife Zulma and daughter Clarissa in Las Cruces, New Mexico, enjoying the peaceful tranquility and scenery of Southern New Mexico.

GLOSSARY

Budget—The amount of funding, resources, and investments allocated to a project.

Business Requirements—The business needs that are the driving force behind the project.

Chief Information Officer/Chief Technology Officer—An IT professional who has obtained the necessary training, experience, competency, degrees, and certifications to perform all aspects of IT management in strategic business leadership. The position usually requires a Master's or Ph.D. degree in a computer-related or business field and usually entails oversight of all IT and associated business units. The holder of this position is usually an executive officer of a company and may directly report to either the Chief Executive Officer or Chief Financial Officer of that company.

Contingency Planning—The process of identifying alternative actions to be taken if an identified risk should occur (i.e., backup plans).

Customer—The individual or organization that will receive the service or hardware generated by the project.

Director of IT—An IT professional who has obtained the necessary training, experience, competency, degrees, and certifications to perform all aspects of IT management, in terms of both operations and tactical business processes. The position usually requires a Bachelor or Master's degree in a computer-related or business field and usually entails oversight of various IT support areas including but not limited to: Help Desk Operations, Technical Support, Applications Support, and Network Management.

Documentation—Information used for reference, and the act of gathering this information (the 5th step of the *NuneX Method* of technical documentation techniques).

End User—The person or people who will be affected by the project or technology users. Often termed "knowledge" workers.

Estimating—The process of determining the costs associated with performing a project for hardware, software, labor, etc.

Execution—The stage in the project where all plans are carried out or executed.

Help Desk—In a business enterprise, a help desk is a place that a user of information technology can call to get help with a problem. In many companies, a help desk is simply one person with a phone number and a more or less organized idea of how to handle the problems that come in. In larger companies, a help desk may consist of a group of experts using software to help track the status of problems, and other special software to help analyze problems (for example, the status of a company's telecommunications network). Typically, the term is used for centralized help to users within an enterprise. A related term is "Call Center," a place that customers call to place orders, track shipments, get help with products, and so forth. The World Wide Web offers the possibility of a new, relatively inexpensive, and effectively standard user interface to help desks (as well as to call centers) and appears to be encouraging more automation in help desk service.

Some common names for a help desk include: Computer Support Center, IT Response Center, Customer Support Center, IT Solutions Center, Resource Center, Information Center, and Technical Support Center.

Information—Relates to description, definition, or perspective (what, who, when, where).

Information Systems (IS)—The use of and investment in computer technology by the principal or centralized organization formally charged with the responsibility for computer technology. Information Systems is also the formal name of the department within an enterprise that is responsible for the Information Technology. Other common names for the IS department are:

— Management Information System or MIS

— Data Processing or DP

— Information Processing or IP

Information Technology (IT)—A term that encompasses all forms of technology used to create, store, exchange, and use information in its various forms (business data, voice conversations, still images, motion pictures, multimedia presentations, and other forms, including those not yet conceived). It's a convenient term for including both telephony and computer technology in the same word. It is the technology that drives what has often been called "the information revolution."

Knowledge—Comprises strategy, practice, method, or approach (how).

Knowledge Management (KM)—KM is the process through which organizations generate value from their intellectual and knowledge-based assets. Most often, generating value from such assets involves sharing them among employees and departments, and even with other companies, in an effort to devise best practices. It's important to note that the definition says nothing about technology. While KM is often facilitated by IT, technology by itself is not KM.

Maintenance—Step 7 of the *NuneX Method* is maintaining and updating your documentation hours, days, weeks, months, years, etc. after the project is completed or problem is resolved, within reason and scope. You want to keep updating the results of your work, as sometimes problems and resolutions have a cause and effect basis that may take time to reveal. Having a recorded accounting of events after you perform the project can help you refer back in case you need to perform clean-up work or analyze data for future reference.

Network Administrator—An IT technical professional who has obtained the necessary training, experience, competency, degrees, and certifications to perform local area network and wide area network service work, repair, troubleshooting, and implementations. The network administrator will often be responsible for the overall performance and maintenance of a company's network infrastructure. The position usually requires a bachelor's degree or the attending of a technical college focused on Information Technology and network management.

Network Engineer—An IT technical professional who has obtained the necessary training, experience, competency, degrees and certifications to perform all aspects of network service work, repair, troubleshooting, design, engineering, and implementations. The network engineer is usually considered one step above the technical level of a network administrator. The position usually requires a bachelor's degree in a computer-related field.

Network Manager—An IT technical professional who has obtained the necessary training, experience, competency, degrees, and certifications to perform all aspects of network management. The network manager usually manages network engineers, network administrators, and/or network technicians. The position usually requires a Bachelor's degree in a computer-related field and oversight of other technical professionals including but not limited to: network administrators, network technicians, and personal computer technicians.

Network Technician—An IT technical professional who has obtained the necessary training, experience, competency, degrees, and certifications to perform local area network service work, repair, troubleshooting, and implementations. The position is usually considered to be technically one step above a Personal Computer Specialist and usually requires an Associate degree or the attending of a technical college focused on information technology.

Notation—The fourth step of the *NuneX Method*. Notes are written information or downloaded documents, diagrams, excerpts from publications, etc. that you may use as references, guides, or troubleshooting aids for your project. This step also requires organizing these resources for your own personal use so you don't have to look up the information as you work.

NuneX Method—The *NuneX Method*, named after its developer, Richard Nuñez, is a 7-Step process of Technical Documentation Techniques for IT professionals who work in technical and engineering level positions. The 7 steps are as follows:

1. Pre-Documentation
2. Planning
3. Security
4. Notation
5. Documenting
6. Refinement
7. Maintenance & Updating

Following these 7 steps will allow an IT professional to achieve greater documentation and allow for better quality service and professional advancement.

Objectives—Predetermined and intended results, outcomes, or deliverables toward which effort is directed on a project.

OSI (Open Systems Interconnection)—*Definition obtained from* http://www.whatis.com. The OSI is a standard description or reference model for how messages should be transmitted between any two points in a telecommunication network. Its purpose is to guide product implementers so that their products will consistently work with other products. The reference model defines seven layers of functions that take place at each end of a communication. Although OSI is not always strictly adhered to in terms of keeping related functions together in a well-defined layer, many if not most products involved in telecommunication make an attempt to describe them in relation to the OSI model. It is also valuable as a single reference view of communication that furnishes everyone a common ground for education and discussion.

Developed by representatives of major computer and telecommunication companies beginning in 1983, OSI was originally intended to be a detailed specification of interfaces. Instead, the committee decided to establish a common reference model for which others could develop detailed interfaces that in turn could become standards. OSI was officially adopted as an international standard by the International Organization of Standards (ISO). Currently, it is Recommendation X.200 of the ITU-TS.

The main idea in OSI is that the process of communication between two end points in a telecommunication network can be divided into layers, with each layer adding its own set of special, related functions. Each communicating user or program is at a computer equipped with these seven layers of function. So, in a given message between users, there will be a flow of data through each layer at one end down through the layers in that computer and, at the other end, when the message arrives, another flow of data up through the layers in the receiving computer and ultimately to the end user or program. The actual programming and hardware that furnishes these seven layers of function is usually a combination of the computer operating system, applications (such as your Web browser), TCP/IP or alternative transport and network protocols, and the software and hardware that enable you to put a signal on one of the lines attached to your computer.

OSI divides telecommunication into seven layers. The layers are in two groups. The upper four layers are used whenever a message passes from or to a user. The lower three layers (up to the network layer) are used when any message passes through the host computer. Messages intended for this computer pass to the upper layers. Messages destined for some other host are not passed up to the upper layers but are forwarded to another host. The seven layers are:

Layer 7: The application layer...This is the layer at which communication partners are identified, quality of service is identified, user authentication and privacy are considered, and any constraints on data syntax are identified. (This layer is *not* the application itself, although some applications may perform application layer functions.)

Layer 6: The presentation layer...This is a layer, usually part of an operating system, that converts incoming and outgoing data from one presentation format to another (for example, from a text stream into a popup window with the newly arrived text). Sometimes called the syntax layer of the OSI.

Layer 5: The session layer...This layer sets up, coordinates, and terminates conversations, exchanges, and dialogs between the applications at each end. It deals with session and connection coordination.

Layer 4: The transport layer...This layer manages the end-to-end control (for example, determining whether all packets have arrived) and error-checking. It ensures complete data transfer.

Layer 3: The network layer...This layer handles the routing of the data (sending it in the right direction to the right destination on outgoing transmissions and receiving incoming transmissions at the packet level). The network layer does routing and forwarding.

Layer 2: The data-link layer...This layer provides synchronization for the physical level and does bit-stuffing for strings of 1's in excess of 5. It furnishes transmission protocol knowledge and management.

Layer 1: The physical layer...This layer conveys the bit stream through the network at the electrical and mechanical level. It provides the hardware means of sending and receiving data on a carrier.

Personal Computer Technician—An IT technical professional who has obtained the necessary training, experience, competency, degrees, and certifications to perform personal computer service work, repair, troubleshooting, and implementations. The position is usually considered to be technically one step above a Help Desk professional and usually requires an Associate degree or the attending of a technical college focused on information technology.

Planning—The 2nd phase in the *NuneX Method* of technical documentation wherein a technical resource plans all activities for a project.

Pre-Documentation—Is the first step of the *NuneX Method*, which involves setting up Pre-Documentation checklists and organizing your approach, materials, and resources. It involves putting together a resource information database that contains names, addresses, phone lists, e-mail, web-site information, etc. It also involves the psychological preparation that one needs to get started on a project.

Project—A request, repair, implementation, or any technical work requiring documentation. According the Project Management Institute or PMI, a project is "a temporary endeavor undertaken to create a unique product or service."

Project Management—An approach used to mange work within the constraints of time, cost, materials, and performance targets.

Project Management Plan—A document that outlines the project in terms of the business need, objectives, and how to achieve the objectives. It uses the scope document as its foundation and addresses items such as reporting guidelines, tracking methods, risk management, and budget management.

Project Manager—The individual responsible for the day-to-day management of a project.

Project Scope—The phase within a project that determines who, what, when, where, why, and how.

Project Status Report—This report documents the status of the design, development efforts, and progress toward achieving the requirements of the project. It includes the measurement of the relationship between the actual progress and the scheduled progress.

Project Team—Those who report to the project manager or departmental management and are key resources involved in the project.

Refinement—Step 6 of the *NuneX Method*, wherein refinement of the documentation that was taken during step 5 is seen to. This step calls upon the documenter to reflect back on his or her documentation in Step 5 and notate what contributed to his or her success and/or failure to produce the desired objectives and deliverables. In a nutshell, the documenter goes back and formally documents what actually worked and what pitfalls are to be avoided the next time. It is the final documentation product and deliverable that a documenter may provide to a customer or employer, or for their own records. The documenter may also notate the training and skill sets that someone else may need in order to assume the Maintenance phase of the project. Knowledge Management encompassing the answers to the questions of who, what, when, where, how, and why are key to this step.

Return on Investment (ROI)—Financial gain expressed as a percentage of funds invested to generate that gain.

Scope Change—Any change within a project that requires a change in the project's objectives, cost, or schedule.

Scope Creep—Is the gradual addition of new requirements to the original specifications of a project. As the requirement list increases, the project's complexity increases, often causing a project to fail.

Security—Is the third step of the *NuneX Method* which requires the use of a special encrypted coding method to securely document passwords for referencing system access. If your documentation falls into the wrong hands, you want to protect your interests and deter hacks from the systems you are held accountable for.

Statement of Work or Scope of Work—Description of the products and services to be obtained under a written contractual agreement between parties.

Success—Is the meeting and/or exceeding of set objectives and goals.

Systems Analyst—An IT professional who has obtained the necessary training, experience, competency, degrees, and certifications to perform all aspects of systems analysis, design, and control. The position usually requires a Bachelor's degree in a computer related or business field and is usually focused on software, operating, database, and process flow control systems.

Systems Engineer—An IT technical professional who has obtained the necessary training, experience, competency, degrees, and certifications to perform all aspects of network service work, repair, troubleshooting, systems design, engineering, and implementations. The network engineer is usually considered one step above the technical level of a network engineer since he or she usually provides technical support and guidance to his or her internal and external customers. The position usually requires a bachelor's degree in a computer related field and is usually part of the reseller, vendor, and manufacturer support structure of IT.

Technical Shorthand (TS)—Technical Shorthand is a system or form of rapid handwriting employing symbols and abbreviations to represent words, phrases, and letters for reference. The *NuneX Method* utilizes TS during the documentation stage for quick and easy documenting of actual events as they occur within a project.

Technical Resource—Can be anyone who is part of an IT department's technical staff or team; a contractor, consultant, etc. who performs technical work.

Wisdom—Embodies principles, insights, morals, or archetypes (why).

0-595-66649-3

www.ingramcontent.com/pod-product-compliance
Lightning Source LLC
LaVergne TN
LVHW042137040326
832903LV00011B/281/J